SpringerBriefs in Philosophy

SpringerBriefs present concise summaries of cutting-edge research and practical applications across a wide spectrum of fields. Featuring compact volumes of 50 to 125 pages, the series covers a range of content from professional to academic. Typical topics might include:

- A timely report of state-of-the art analytical techniques
- A bridge between new research results, as published in journal articles, and a contextual literature review
- A snapshot of a hot or emerging topic
- An in-depth case study or clinical example
- A presentation of core concepts that students must understand in order to make independent contributions

SpringerBriefs in Philosophy cover a broad range of philosophical fields including: Philosophy of Science, Logic, Non-Western Thinking and Western Philosophy. We also consider biographies, full or partial, of key thinkers and pioneers.

SpringerBriefs are characterised by fast, global electronic dissemination, standard publishing contracts, standardised manuscript preparation and formatting guidelines, and expedited production schedules. Both solicited and unsolicited manuscripts are considered for publication in the SpringerBriefs in Philosophy series. Potential authors are warmly invited to complete and submit the Briefs Author Proposal form. All projects will be submitted to editorial review by external advisors.

SpringerBriefs are characterised by expedited production schedules with the aim for publication 8 to 12 weeks after acceptance and fast, global electronic dissemination through our online platform SpringerLink. The standard concise author contracts guarantee that

- an individual ISBN is assigned to each manuscript
- each manuscript is copyrighted in the name of the author
- the author retains the right to post the pre-publication version on his/her website or that of his/her institution.

Henrik Vogt
Department of Community Medicine
and Global Health
Institute of Health and Society, University
of Oslo
Oslo, Norway

Maxence Gaillard
Centre for Medical Ethics
Institute of Health and Society, University
of Oslo
Oslo, Norway

Sara Green
Section for History and Philosophy
of Science, Department of Science
Education
University of Copenhagen
Copenhagen, Denmark

ISSN 2211-4548   ISSN 2211-4556 (electronic)
SpringerBriefs in Philosophy
ISBN 978-3-031-85534-4   ISBN 978-3-031-85535-1 (eBook)
https://doi.org/10.1007/978-3-031-85535-1

© The Editor(s) (if applicable) and The Author(s) 2025. This book is an open access publication.

Open Access This book is licensed under the terms of the Creative Commons Attribution 4.0 International License (http://creativecommons.org/licenses/by/4.0/), which permits use, sharing, adaptation, distribution and reproduction in any medium or format, as long as you give appropriate credit to the original author(s) and the source, provide a link to the Creative Commons license and indicate if changes were made.
The images or other third party material in this book are included in the book's Creative Commons license, unless indicated otherwise in a credit line to the material. If material is not included in the book's Creative Commons license and your intended use is not permitted by statutory regulation or exceeds the permitted use, you will need to obtain permission directly from the copyright holder.
The use of general descriptive names, registered names, trademarks, service marks, etc. in this publication does not imply, even in the absence of a specific statement, that such names are exempt from the relevant protective laws and regulations and therefore free for general use.
The publisher, the authors and the editors are safe to assume that the advice and information in this book are believed to be true and accurate at the date of publication. Neither the publisher nor the authors or the editors give a warranty, expressed or implied, with respect to the material contained herein or for any errors or omissions that may have been made. The publisher remains neutral with regard to jurisdictional claims in published maps and institutional affiliations.

This Springer imprint is published by the registered company Springer Nature Switzerland AG
The registered company address is: Gewerbestrasse 11, 6330 Cham, Switzerland

If disposing of this product, please recycle the paper.

Henrik Vogt · Maxence Gaillard · Sara Green

Organoids for Personalised Cancer Medicine

A Vision Assessment of an Emerging Health Technology

Preface and Acknowledgments

The work presented in this book was conducted in the context of the HYBRIDA project. HYBRIDA—*Embedding a comprehensive ethical dimension to organoid-based research and relating technologies*—was funded through the H2020 Science with and for Society programme (grant agreement 101006012). The main outputs of HYBRIDA are operational guidelines for organoid research and contributions to ethics and normative frameworks. As a groundwork, HYBRIDA explored different forms of uncertainty pertaining to organoid research: ontological, conceptual, epistemological, ethical and regulatory aspects, as well as public attitudes.

This book springs out of the Work Package 2 (WP2) of HYBRIDA, which was dedicated to epistemological uncertainty pertaining to these models, and especially to their use in medicine. In this work package, a mapping of the organoid field was first conducted (see Shoji et al. 2023), and then a traditional health technology assessment (HTA) looking at evidence from randomised controlled trials (see Hofmann et al. 2023).

This book represents a condensed and reworked version of two already published HYBRIDA reports: D2.3—Adaptation of health technology assessment (HTA) to evaluate organoids and organ-on-a-chip as emerging technologies in the clinic and D2.4–An amended health technology assessment (HTA) to evaluate organoids as emerging technologies in the clinic. They may be accessed on HYBRIDA's website[1] for more detail. The book seeks to present the results in a readable format for an audience of scientifically oriented philosophers, philosophers interested in biomedical science (particularly organoids or precision medicine), philosophically oriented scientists, as well as readers interested in technology assessment of emerging technologies and visions. The reports had a broad scope in covering the use of organoids and organ-on-a-chip technology in the study and management of a variety of diseases, particularly cystic fibrosis. This book narrows the scope to focus on the frontier clinical application of organoids in precision oncology.

---

[1] https://hybrida-project.eu/.

At the time of research, Henrik Vogt (HV) was HYBRIDA WP2 leader, Maxence Gaillard (MG) and Sara Green (SG) were associated researchers and main contributors to WP2. Part of the work behind this book was done by HV in the context of a postdoctoral fellowship at the Hybrid Technology Hub, Centre for Organ on a Chip-Technology, University of Oslo. It was funded by the Norwegian Research Council, grant/award Number 262613.

We would like to thank all our colleagues and mentors from HYBRIDA: Jan Helge Solbakk who was the initiator and the soul of the project until he stepped down as PI, Søren Holm who took over as PI, Bjørn Hofmann who was the previous WP2 leader, and Panagiotis Kavouras, who has been a central organising force in HYBRIDA at all stages. Our work benefited from formal and informal input from many HYBRIDA members, especially Christine Mummery, Richard Davies, Stefan Krauss and Junya Shoji. Panagiotis Kavouras, Jonathan Lewis, and Søren Holm were members of a working group that contributed to delineating the workplan for this research. Panagiotis Kavouras has been involved in the WP2 interview task and contributed to transcription. Jonathan Lewis was involved in the development of the registered trials material. The focus groups conducted by Tine Ravn and colleagues provided insights on hype and communication. Panagiotis Kavouras, Jonathan Lewis, Richard Davies and external experts Martin Sand, Mary Walker, Megan Munsie (the latter also a HYBRIDA advisory board member) reviewed parts of our HYBRIDA reports and provided very helpful comments.

Last, but not least, an important element of our research is interviews with experts in the field at "epistemological hotspots", and we would like to thank them for their time and very enlightening conversations.

Oslo, Norway
Henrik Vogt
Maxence Gaillard
Sara Green

References

Hofmann B, Zinöcker S, Holm S, Lewis J, Kavouras Pa (2023) Organoids in the clinic: A systematic review of outcomes. Cells Tissues Organs 212(6):499–511. https://doi.org/10.1159/000527237

Shoji J, Davis R, Mummery, CL, Krauss S (2023) Global meta-analysis of organoid and organ-on-chip research. Adv Healthc Mater no 2301067. https://doi.org/10.1002/adhm.202301067

Contents

1 Introduction 1
 1.1 Organoid Technology and Its Promises 2
 1.2 Epistemological Issues in Precision Medicine 5
 1.3 Assessing Emerging Technologies and Visions 7
 1.4 Consent and Ethics Approval 8
 References 8

2 Vision Assessment as a Philosophically Based Health Technology Assessment for Emerging Technologies 11
 2.1 Technology Assessment and Health Technology Assessment 11
 2.2 Vision Assessment 13
 2.3 Other Relevant Approaches to Assess and Guide Technology Development 17
 2.4 Material and Methods 19
  2.4.1 Development and Analysis of Review Material 19
  2.4.2 Development and Analysis of Epistemological Hotspots Material 21
  2.4.3 Development and Analysis of Material of Published Clinically Relevant Evidence 22
  2.4.4 Development and Analysis of Material of Registered, Clinically Relevant Trials 24
  2.4.5 Further Development of Vision Analysis and Vision Evaluation 25
 References 27

3 Vision Analysis of Patient-Derived Organoids for Personalised Medicine 29
 3.1 The Overarching Promise: Improved Predictions and Clinical Utility Through Personalisation 29
  3.1.1 Different Stem Cells, Different Aspects of the Promise 31
  3.1.2 A Promise of Precision Oncology 32

		3.1.3	Different Degrees of Boldness in the Vision	34
	3.2	\multicolumn{2}{l}{Main Expectation A: Improved Prediction Through More Representative Models}	35

  3.1.3 Different Degrees of Boldness in the Vision 34
 3.2 Main Expectation A: Improved Prediction Through More
  Representative Models 35
  3.2.1 The Assumption of Genetic Representation 36
  3.2.2 The Microenvironment—A Promise of Context
   and Holism 37
  3.2.3 The Assumption of Self-organisation as a Design
   Principle 38
 3.3 Main Expectation B: Practical Feasibility 38
  3.3.1 Functional Precision Medicine, Functional
   Biomarkers, and Molecular Agnosticism 38
  3.3.2 The Clinical Setup: How Personalised Medicine is
   Expected to Be Delivered 39
  3.3.3 Prerequisites and Conditions for Practical Feasibility 43
 3.4 Main Expectation C: Amenability and Documentation 44
  3.4.1 Observational Trials (Parallel Trials and Co-clinical
   Trials) 44
  3.4.2 Interventional Trials—"Prospective Validation Trials"
   and Randomised Controlled Trials 45
  3.4.3 Single-Case Design Trials and Case Series 46
 References 48

4 Vision Evaluation of Patient-Derived Organoids for Personalised Medicine 53
 4.1 Uncertainty Regarding Main Expectation A: Representation
  (Model Uncertainty) 53
  4.1.1 Issues with the Harvesting and Processing of Cells 54
  4.1.2 Issues in Accounting for the Environment 55
  4.1.3 Self-organisation: Issues with the Main Design
   Principle 58
  4.1.4 Issues with Size, Maturation and Long-Term Culture 59
  4.1.5 Towards More Holistic Organoid Models of Higher
   Complexity 60
  4.1.6 Simplicity Versus Complexity: The
   Accuracy-Reliability Trade-Off 61
 4.2 Uncertainty Regarding Main Expectation B: Practical
  Feasibility 62
  4.2.1 Lack of Standardisation and Technical Noise 62
  4.2.2 Establishment Success Rate: Problems
   with Generating Organoids 65
  4.2.3 Generating Enough Organoids for Testing 66
  4.2.4 "At the Right Time"? Challenges of Speed in Growth 67
  4.2.5 Gaining the Trust of Clinicians and Evidence-Based
   Medicine as an Obstacle 67

		4.2.6	Biobanks and the Stratified Approach as a Solution to Technical and Practical Issues	68
	4.3	Uncertainty Regarding Main Expectation C: Amenability and Documentation		69
		4.3.1	Summary of Published Clinically Relevant Evidence	70
		4.3.2	Discussion and Conclusion: The Evidence Status of the Vision	72
		4.3.3	Summary of Registered Trials: The Translational Pipeline	75
		4.3.4	Discussion and Conclusion: Registered Trials and the "Translational Pipeline"	78
		4.3.5	Hurdles: "Standardisation, Standardisation, Standardisation"	80
		4.3.6	The Tension Between Standardisation and Personalisation	82
		4.3.7	N-of-1 Situations: Knowing What Works in the Unique Case	84
	4.4	Conclusion of Vision Evaluation		87
		4.4.1	Credibility and Hype	88
		4.4.2	Utility in Context	89
		4.4.3	Problems of Knowing and Documenting that It Works	90
	4.5	Limitations of Our Method		90
	References			91
5	**Vision Management: Towards Responsible Visioneering for Tumour Organoid Technology in Precision Oncology**			**97**
	5.1	Summary of Results		97
	5.2	Does the Community Correctly Know the Evidence? Addressing Publication Bias		98
	5.3	How Long Until Clinical Application? Striving for Standardisation		99
	5.4	Who Would Benefit? Addressing Economics and Distributive Justice Issues		100
	5.5	How Should One Communicate the Vision? Dealing with Hype		102
	References			104
6	**Summary and Concluding Remarks**			**105**
Appendices				**109**
Index				**125**

Chapter 1
Introduction

[V]isions have two major features: they are a mental image of an attainable future shared by a collection of actors; and they guide the actions of and interactions between those actors.

John Grin (2000)

This book is intended for anyone interested in the careful evaluation of the impact of organoid technology on future medicine and for readers interested in how an evaluation of emerging technologies can be done more generally. Hence, in this book, we seek to provide the reader with two things: First, an examination of a particular vision for the future of medicine. Second, since no off-the-shelf method exists for an analysis of such visions, we will contribute to the development of a philosophically based strategy for examining such intangible objects of the future. The vision we will scrutinise is the idea of using organoids for personalised cancer medicine. We will analyse its plausibility, its practical feasibility, and touch upon its responsibility. Concerning the latter part of this undertaking, we will show that such an analysis can meaningfully be done and add new elements to an already existing framework: vision assessment.

Concerning our vision of interest, we will tease apart the different assumptions, prerequisites, and preconditions that it is built upon, and show how several of them have shaky foundations. The key concept in this book is *uncertainty*. There are considerable epistemological uncertainties pertaining to the promises made about organoids, which we will be identifying and discussing. One form of uncertainty we will highlight is *model uncertainty*. There are significant differences between the organoids as models of reality and the reality they model—with significant consequences for their potential. Some of these differences are not just about technical problems and the current state of the models—but about more fundamental limitations in how in vitro models can be built and in our very knowledge about biology. We will also highlight what technical challenges and uncertainties must be overcome to move the field forward towards clinical application, as well as ethical challenges associated with their evidence status, including concerns about publication bias.

© The Author(s) 2025
H. Vogt et al., *Organoids for Personalised Cancer Medicine*,
SpringerBriefs in Philosophy, https://doi.org/10.1007/978-3-031-85535-1_1

Science and medicine need visions to guide our present actions and investments, and they should not be constrained to be easily realisable. We will argue that, while the bolder versions of the organoids for cancer medicine vision are hyped and unrealistic, its more modest claims seem justified.

Before delving into the vision and reaching our conclusions, we will first introduce the reader to our subject and objectives. We then develop the method for our undertaking and apply this to the emerging technology of organoids as personal models in cancer medicine.

Throughout the book, the reader may choose to be selective and skip parts that are not perceived as relevant. Those interested in methodological development may focus on this strand in its contents, those interested in organoids and personalised medicine may follow the latter more closely.

1.1 Organoid Technology and Its Promises

New technologies, enabling humans to do things that were previously impossible, may be the closest thing a person adhering to the tenets of naturalism can come to magic (Clarke 1962). Technologies and scientific breakthroughs can inspire great visions and promises about what the future holds, and some more than others. In recent years, many great visions have directed humanity to work inwards, into our bodies in search of means to cure human disease and prolong life. We have seen the fast development of several biotechnologies, and among the most well-known ideas is that of developing precision medicine or personalised medicine (hereafter abbreviated PM). PM efforts are animated by the dream of "delivering the right treatments, at the right time, every time to the right person" (White House 2015). Recently, a new technology has been promoted for this purpose: organoids. The word organoid itself means an "organ-like" entity. This ability to be *something like* the body's own systems, while still being outside the body (that is, in vitro) and open to experiments is what makes organoids useful. It is the latest development in humanity's quest to find something *like ourselves* to experiment on, as we cannot, for ethical and practical reasons, experiment on human beings. The focus of this book is about the convergence of two visions: It is about the way in which the vision of organoids shapes the PM vision, and how the resulting promises can be assessed. To be more precise, organoids should not be viewed as one tool, but rather as a family of biotechnologies emerging in a new wave since approximately 2010. Neither should they be regarded as entirely novel. Biomedical research has relied on in vitro methods for decades, but techniques for culturing, or growing, cells have become increasingly advanced (Simian and Bissell 2017), and organoids were labelled "method of the year" by *Nature Methods* in 2017 (De Souza 2018). Experimental research in this area has provided novel insights to factors influencing how stem cells differentiate to become other, more specialised cells, tissues, and organs. This has allowed scientists to take stem cells and develop three-dimensional (3D) constructs that reproduce certain anatomical and physiological features of the body in a dish.

1.1 Organoid Technology and Its Promises

In landmark achievements around 2010, researchers established intestinal organoids—small structures resembling pieces of gut—first in mice, then in humans (Sato et al. 2009; Jung et al. 2011). Since then, many kinds of organoids have been developed, from liver, pancreas, or kidney to neural organoids. Building on the prowess in the early 2010s, organoids technology is getting closer to more tangible, large-scale delivery, with the institutionalisation of biobanks, growing industrial interest, and a range of applications and promises in different settings. Bose et al. (2021, p. 1011), for example, write that:

> Organoids are self-organizing, expanding 3D cultures derived from stem cells. Using tissue derived from patients, these miniaturized models recapitulate various aspects of patient physiology and disease phenotypes including genetic profiles and drug sensitivities. As such, patient-derived organoid (PDO) platforms provide an unprecedented opportunity for improving preclinical drug discovery, clinical trial validation, and ultimately patient care.

At first, organoids are a valuable tool for basic research in developmental biology. They open avenues to observe and understand key mechanisms of embryonic development in vitro. Another important application is to understand the mechanisms behind disease through *disease modelling* (Clevers 2016). Organoids can also be controlled and manipulated (e.g., by gene editing) to make them more useful for specific purposes in basic science. Beyond basic research, organoids are already competing with and complementing animal models in drug development, testing and toxicity screening, which are essential steps in preclinical research (Walsh et al. 2014). In this book, however, we have chosen to focus only on their promised role as clinical applications.

In the clinic, in the real-life practice of medicine by doctors, organoids are envisioned to be used directly in the management of patients. One avenue for using organoids clinically is transplantation or implants for regenerative medicine, including implants developed from the patient's own cells. While regenerative medicine is a promising area of future applications, we here focus on a different idea: To use organoids as predictive models to diagnose and personalise treatments for individual patients. The aim is to know just what a person needs, by performing tests on an organoid that mimics that person's disease.

We have chosen to focus on precision oncology (cancer medicine) because the prospect of improving cancer care features prominently in the vision. Moreover, alongside cystic fibrosis, a rare genetic disease, cancer has progressed the furthest towards the clinic in the organoid field. During the first steps of our inquiry, we also studied organ-on-chip models, which are more engineered in that they combine cell cultures with chips that for example contain pumps and valves to mimic blood vessels. We learned that, although related, organ-on-a-chip research has different challenges and is generally further from clinical implementation. We will thus touch upon organ-on-a-chip models only when relevant to our main objective.

The focus of this book is a specific kind of organoid called *patient-derived organoids (PDOs)*. They are grown from cells harvested from human patients. When they are derived from a patient's tumour or cancer cells and turned into a model of that person's cancer, they may be called *patient-derived tumour organoids (PDTOs)*.

The focus of this book is the vision of using the latter kind of organoid in the clinic to tailor treatments to specific cancer patients in what is known as *precision oncology* (PO). We will label this the *patient-derived tumour organoid for precision oncology vision* (*"PDTO for PO"*). These models are expected to mimic the patient's specific disease and, when tested with different drugs, to predict treatment response in that individual. In this book, we analyse and evaluate (critically assess) this vision.

Organoid Ethics

As organoids gained momentum in biomedical research, this technology has become a hot topic in the bioethics literature as well (Bredenoord et al. 2017; Barnhart and Dierickx 2022), with most discussions focusing on the best ethical framework or regulation for the field (e.g., ISSCR 2021). Some applications of organoid technology have drawn more attention than others. This is especially the case for neural organoids and embryo models, entities that raise manifest ethical concerns when grown in a dish. How should we consider neural organoids if they might become sentient at some point in the future (Sawai et al. 2021)? How should we consider stem cell-based embryo models, which look more and more like fertilisation embryos (Rivron et al. 2023)? In both cases, the normative issue revolves around the moral status of the entity of concern: the entity obtained from stem cells acquires some properties that force us to Reconsider its status, as it would be unethical to relate to this entity as a regular cell culture. Other issues extensively discussed in the bioethics literature include biobanking and informed consent, on the grounds that it is difficult to anticipate what the collected cells will become and how they will be transformed in the future. This is still, in a way, putting the focus on the new entity to be developed via organoid technology and asking how special they are.

By contrast, tumour organoids have drawn much less attention. This might be because tumours have no specific moral status: Aside from immortalised cell lines used in research, such as HeLa cells (Landecker 2007), material from biopsies or tumour resections is mostly treated as surgical waste and discarded or used as material with purely instrumental value (e.g., for pathological analysis). Unlike organoids that look like "interesting" organs, tumour organoids do not raise the same interest for themselves. While neural organoids and embryo models raise specific ethical concerns due to their developmental potential, tumour organoids are not associated with the same inherent moral status and have mostly instrumental value as long as they serve the interests of the persons behind them. Nonetheless, we show here that tumouroids can raise other ethical issues related to epistemological uncertainty in personalised medicine that bioethics should address as well.

1.2 Epistemological Issues in Precision Medicine

Another way of phrasing the aim of this book, is to say that we want to identify *epistemological uncertainty* pertaining to the realisation of the vision. Epistemology is the philosophical study of knowledge and limits to knowledge. It underpins methodology, how research is designed, and how inferences are made. It also underpins visions; what researchers think they can know about the future and achieve accordingly. By exploring epistemological issues in this book, we examine what is known and what is not about these technologies and evaluate the premises on which the organoid vision for precision oncology is built.

In this story, the concept of *evidence-based medicine* (EBM) plays a dual role: First, because it is what the idea of personalised or precision medicine is often compared or put in opposition to, and second, because EBM forms the backbone of how technologies, emerging or established, are evaluated in the field of health technology assessment (HTA).

When an innovation comes close to the clinic, it is usually tested for safety, efficacy and effectiveness before it might become adopted. For decades, our healthcare systems have relied on EBM as epistemological paradigm (Guyatt et al. 2015). A hallmark of EBM is high numbers: If a treatment is to be considered successful, it will be after a collection of results covering many patients and situations, including controlled clinical trials with human patients, resulting in quantitative analysis with scores passing determinate statistical thresholds. The randomised clinical trial (RCT), a population-based, controlled experiment with an intervention and control group, is considered the "gold standard" test that a treatment should pass. Clinical trials follow strict procedures from registration to the analysis of results, so that the confidence we can have in this new treatment, drug, or innovation is based on the observation of its efficiency documented through systematic comparison. EBM does not completely disregard other kinds of evidence, like observations in a limited series of patients or even single patients. However, a hierarchy of evidence (from low reliability single cases to robust results established in RCTs) serves as a guide when it comes to compiling existing evidence and determining whether something new deserves to enter clinical practice (Guyatt et al. 2011). Indeed, EBM is not only a scientific enterprise in the sense that it belongs to academia, in which all fields of research set their own standards. EBM is a set of practices and rules at the political and regulatory level that contribute to determining which treatments are legitimate and which are not. Market authorisations and reimbursement procedures rely for an important part on EBM and the scientific evaluation of innovations with big numbers. When it comes to the decision process of letting a drug enter the market, EBM is not "applying pure science to the art of medicine" but the name of an entanglement of administrative procedures, policy decisions, and market forces. In other words, there is a lot of power attached to setting the standard for how a treatment can be deemed to work. The very idea of what constitutes "good enough" evidence in medicine opens and shuts doors for various agents, including patients who want to access treatment.

Since it matters so much, there are debates about what EBM should entail. Philosophers of medicine and many other observers have critically examined the conditions for inferences about treatment efficacy and safety issues. Standards of proof always have some limitations, and statistics can be a dangerous tool if it leads to overconfidence in disputable results (Ioannidis 2005; Stegenga 2018a). Also, and maybe more importantly for our topic, treatments that are statistically effective at the level of a population of patients might not work for the patient at hand. Or, as philosophers of medicine put it, there is always uncertainty in applying class (population) probability to case probability (Djulbegovic et al. 2011; Deaton and Cartwright 2018). In that sense, EBM will endorse medicines that are efficient on average as many patients will experience an effect, but this may also leave many individuals without treatment or even harmed. This is especially true in cancer medicine, where many individual patients will have tumours that will resist treatments that have been shown to work at a global scale. This is a problem for these patients, both because of side-effects and potentially in terms of lost opportunity of another alternative treatment. For society and individuals, this also presents an economic problem of wasted resources.

In this context, precision medicine (PM) is often presented as a solution. For cancer, one of the strategies that can be deployed is adapting the treatment (e.g., chemotherapy) according to the genetic profile of the tumour. We know that some tumours resist treatment, but if we knew which mutation indicates which tumours are more susceptible to be targeted by treatment X, and which mutation by treatment Y, then we would enter an era of precision medicine.

PM may be seen as a challenge to the evidence standards traditionally favoured in EBM. It is presented as an alternative to EBM or as altering its contents. Issues arise because PM ultimately aims to diagnose and treat the individual. This means that the number of people deemed to have a condition (n), and thus on whom research can be conducted, converges towards one ("n = 1"). This may leave statistically based methods, like randomised controlled trials, hard to achieve. They may also be rendered less relevant, if the aim really is to study treatment effects specific to rare cases or even individual patients. Will the PM approach work? A positive line of thinking about PM highlights that, even though there are fewer patients in experiments, one would still be able to show that treatments work because they would have higher efficacy in that smaller number (Stegenga 2018b, p. 157; see also Andreoletti 2018). To evaluate treatment efficacy in this context, however, may inaugurate a form of medicine where one openly accepts more reliance on physiological models, case histories and n-of-1 trials (Tonelli and Shirts 2017; Vogt and Hofmann 2022). Can such a medicine have an acceptable epistemological foundation?

To conclude, in this area, where n = 1 (or very few), EBM may need a new footing, i.e., new forms of knowledge production and ways of handling uncertainty. In this context, organoids thus present a philosophically intriguing new approach to evidence in medicine.

1.3 Assessing Emerging Technologies and Visions

In this book we are going to assess the prospects of organoids as this technology and its promises become entangled with precision medicine. How to proceed?

Generally, in medicine, evaluating the impacts, benefits and harms, of new health technologies is the purpose of health technology assessment (HTA)—"the systematic evaluation of properties, effects, or other impacts of health technology"—with the main purpose of informing policymaking and decision-making in a wide sense (Goodmann 2014). HTA, like technology assessment more generally before it, sprang out of an awareness that technologies can have profound, unintended, and unforeseen consequences on different levels. Practitioners of HTA have also recognised that HTA methodology needs to be flexible and tailored to the technology at hand. However, in practice the focus of HTA has been very much defined by the concern for burgeoning healthcare costs and cost-effectiveness, as well as evidence of safety, efficacy and effectiveness before approval (Oortwijn and Sampietro-Colom 2022). For this purpose, a dominant strand of HTA practice has its most important roots in health economics, focusing mostly on quantifiable evidence and less on factors that cannot be easily quantified, including qualitative evidence and uncertainty, ethics and social issues (Palm and Hansson 2006). It is closely linked to clinical epidemiology, that is, EBM, focusing on reviewing and assessing quantitative evidence, particularly RCTs, systematic reviews and meta-analyses.

This evidence that "traditional HTA" is after, has one important characteristic: *It needs to already exist*. This presents a problem. Organoids are representative of new and emerging science and technologies, also referred to as NESTs (Swierstra and Rip 2007). NESTs are special in that they to a large degree only exist as expectations, promises or visions about the future. While still under development, they are expected to produce a strong impact on society by solving existing problems. There is a lot of power in visions. They are movers of popular, political, and economic support. Thus, while producing this impact they have little existing evidence attached to them. Therefore, they are not easily evaluated with a standard evidence-based HTA approach.

Consequently, what is needed—and what we are after here—is not only an HTA of the utility of organoids based on what evidence may exist, but also an assessment of the vision and its main epistemological assumptions, prerequisites, and conditions. This is a way to do justice to the efforts of the actors in the field.

A note of caution: Bioethicists often jump to conclusions, asking: What would be the consequences if the technology were working? This attitude has been called "*speculative ethics*" (Nordmann 2007) and criticised for uncritically accepting visions and their premises as something plausible, thereby validating them. In doing so, the ethicist might turn something abstract and weakly grounded into something that looks concrete, thus fuelling a possible technological hype.

We do not want to fall into this trap. We do not take for granted what the reiterated prospects are, or which temporary "technological obstacles" are likely to be overcome in the coming years. We want to reconstruct the discourse on organoid technology for PM as precisely as possible and examine the assumptions that make this discourse possible and (im)plausible. In this book, you will not find an answer to the question of whether insurances should reimburse organoid-based therapy or organoid-based therapeutic decisions, but we try to foster critical thinking that deals with this epistemological uncertainty the field is facing.

Thus, we hope that this book will also be interesting for readers interested more broadly in method development for the evaluation of emerging technologies and questions about what constitutes evidence in the context of PM.

Since no standard method for assessing issues of evidence and uncertainty pertaining to a biomedical vision of the future exists, it will first need to be developed, which is done in Chapter 2. Chapter 3 describes the most prominent vision of the potential clinical applications of organoid technology, the "patient-derived cancer organoid for precision oncology" vision (PDTOs for PO). Chapter 4 evaluates this vision by confronting it to the fundamental epistemological uncertainties that we unfold. Chapter 5 draws some normative and ethical conclusions from the inquiry.

1.4 Consent and Ethics Approval

As part of the array of methods employed in our vision assessment, we have conducted interviews with experts in the field (see Sect. 2.4). Potential interviewees were invited by direct email. The participants gave their informed consent on participating in the interview and on it being recorded. Approval from the Norwegian Centre for Research Data was obtained prior to this process (04.10.2022, reference number 517887) and recordings and non-anonymised transcripts were stored in TSD (Service for sensitive data, University of Oslo).

Before circulation, interview transcriptions were anonymised as far as possible, which means that some passages were cut, including all personal anecdotes and references that could help identify the speaker. However, given the specificity of the research enterprise and of the personal expertise, it is very difficult to guarantee full anonymity unless we remove all the interesting information (type of work going on, kind of patients/pathologies that are dealt with, context of the research, etc.). We tried to find a balance between keeping meaningful information and preserving the anonymity of interviewees. Because it might be possible to identify the informants even if we pseudonymise the personal information, we rely here on many quotes from the interviews but the transcription files that provide the background for the interview quotes are not made readily available.

References

Andreoletti M (2018) More than one way to measure? A casuistic approach to cancer clinical trials. Perspect Biol Med 61(2):174–190

References

Barnhart A, Dierickx K (2022) The many moral matters of organoid models: a systematic review of reasons. Med Health Care Philos 25(3):545–560. https://doi.org/10.1007/s11019-022-10082-3

Bose S, Clevers H, Shen X (2021) Promises and challenges of organoid-guided precision medicine. Medicine 2(9):1011–1026. https://doi.org/10.1016/j.medj.2021.08.005

Bredenoord A, Clevers H, Knoblich J (2017) Human tissues in a dish: the research and ethical implications of organoid technology. Science 355:eaaf9414

Clarke A (1962) Profiles of the future: an inquiry into the limits of the possible. Harper & Row, New York

Clevers H (2016) Modeling development and disease with organoids. Cell 165(7):1586–1597

De Souza N (2018) Organoids. Nat Methods 15(1):23

Deaton A, Cartwright N (2018) Understanding and misunderstanding randomized controlled trials. Soc Sci Med 210:2–21. https://doi.org/10.1016/j.socscimed.2017.12.005

Djulbegovic B, Hozo I, Greenland S (2011) Uncertainty in clinical medicine. In: Gifford F (ed) Philosophy of medicine, vol 16. Elsevier, Amsterdam.

Goodman C (2014) HTA 101: National Information Center on Health Services Research and Health Care Technology (NICHSR). Available from: https://www.nlm.nih.gov/nichsr/hta101/

Grin J (2000) Introduction: vision assessment to support shaping 21th century society? Technology assessment as a tool for political judgement. In: Grin J, Grunwald A (eds) Vision assessment: shaping technology in 21st century society towards a repertoire for technology assessment. Springer, Berlin, pp 9–30

Guyatt G et al (2011) GRADE guidelines: 1. Introduction—GRADE evidence profiles and summary of findings tables. J Clin Epidemiol 64(4):383–394. https://doi.org/10.1016/j.jclinepi.2010.04.026

Guyatt G, Rennie D, Meade M, Cook D (eds) (2015) Users' guides to the medical literature: a manual for evidence-based clinical practice, 3rd edn. McGraw-Hill Education

Ioannidis J (2005) Why most published research findings are false. PLoS Med 2(8):e124

ISSCR (2021) International Society for Stem Cell Research guidelines for stem cell research and clinical translation. https://www.isscr.org/policy/guidelines-for-stem-cell-research-and-clinical-translation

Jung P et al (2011) Isolation and in vitro expansion of human colonic stem cells. Nat Med 17:1225–1227

Landecker H (2007) Culturing life, how cells became technologies. Harvard University Press. Cambridge, MA

Nordmann A (2007) If and then. A critique of speculative nanoethics. NanoEthics 1(1):31–46. https://doi.org/10.1007/s11569-007-0007-6

Oortwijn W, Sampietro-Colom L (2022) The VALIDATE handbook—an approach on the integration of values in doing assessments of health technologies. Radboud University Press

Palm E, Hansson SO (2006) The case for ethical technology assessment. Technol Forecast Soc Chang 73:543–558

Rivron N et al (2023) An ethical framework for human embryology with embryo models. Cell 186(17):3548–3557. https://doi.org/10.1016/j.cell.2023.07.028

Sato T et al (2009) Single Lgr5 stem cells build crypt-villus structures in vitro without a mesenchymal niche. Nature 459(7244):262–265. https://doi.org/10.1038/nature07935

Sawai T et al (2021) Mapping the ethical issues of brain organoid research and application. AJOB Neurosci 13:1–14

Simian M, Bissell M (2017) Organoids: a historical perspective of thinking in three dimensions. J Cell Biol 216(1):31–40. https://doi.org/10.1083/jcb.201610056

Stegenga J (2018a) Medical Nihilism. Oxford University Press, Oxford

Stegenga J (2018b) Care and cure: an introduction to the philosophy of medicine. University of Chicago Press, Chicago

Swierstra T, Rip A (2007) Nano-ethics as NEST-ethics: patterns of moral argumentation about new and emerging science and technology. NanoEthics 1:3–20

Tonelli MR, Shirts BH (2017) Knowledge for precision medicine: mechanistic reasoning and methodological pluralism. JAMA 318(17):1649–1650

Vogt H, Hofmann B (2022) How precision medicine changes medical epistemology: a formative case from Norway. J Eval Clin Pract 28(6):1205–1212

Walsh A et al (2014) Quantitative optical imaging of primary tumour organoid metabolism predicts drug response in breast cancer. Can Res 74:5184–5194

White House (2015) Remarks by the President on precision medicine. Available at: https://obamawhitehouse.archives.gov/the-press-office/2015/01/30/remarks-president-precision-medicine

Open Access This chapter is licensed under the terms of the Creative Commons Attribution 4.0 International License (http://creativecommons.org/licenses/by/4.0/), which permits use, sharing, adaptation, distribution and reproduction in any medium or format, as long as you give appropriate credit to the original author(s) and the source, provide a link to the Creative Commons license and indicate if changes were made.

The images or other third party material in this chapter are included in the chapter's Creative Commons license, unless indicated otherwise in a credit line to the material. If material is not included in the chapter's Creative Commons license and your intended use is not permitted by statutory regulation or exceeds the permitted use, you will need to obtain permission directly from the copyright holder.

Chapter 2
Vision Assessment as a Philosophically Based Health Technology Assessment for Emerging Technologies

Where to start when assessing something as nebulous as technologies that are only emerging or that exist as visions? In the first phase of the work leading up to this book, we decided that the first thing we should do was to map the theoretical and methodological literature that has already considered the problem (Vogt et al. 2022). We will here present a condensed version of this review.

Many scholars and other agents have recognised the importance and power of visions and expectations in shaping our world. Consequently, several frameworks or toolkits have explicitly been proposed for the technology assessment or health technology assessment methodologies for such visions. Other frameworks and publications, although not explicitly cast as TA or HTA, also pursue many of the same objectives. They are tied to different fields, such as ethics, sociology, and business.

As we have included this review of different publications on how to engage critically with visions and emerging technologies for readers interested in the methodology of technology assessment, this is a chapter that readers most interested in our assessment of organoids per se can skip.

2.1 Technology Assessment and Health Technology Assessment

Endeavours explicitly called technology assessment (TA) began in the 1960s. They were related to the US Congress wishing to have early warnings and control of technology development. It became more common in Europe in the 1980s (Nazarko 2017). TA was from the beginning directed at trying to predict the future. Early attempts to take an upstream view to predict and assess the influence of technology at the early stages of development have been described as an "expertocratic, positivist, and predictivist conception of TA" (Urueña 2021). The degree to which future consequences can actually be predicted at an early stage is a key point of contention in discussions about how to assess visions. The aims of early TA were soon challenged as too optimistic. This criticism gave rise to alternative methodologies such as "participatory" or "constructive" TA, which appeared in Denmark and the Netherlands in the 1980s, with the main innovation being stakeholder involvement (see e.g. Palm and Hansson 2006). Instead of expecting experts to predict what would actually happen, at least everybody could have their concerns voiced.

However, the quest to predict the future, or at least saying something meaningful about visions, has not been entirely abandoned either. Futures Studies and Foresight have roots in the 1970s and 1980s (Voros 2001). It may be defined as a set of strategic tools "for anticipating fundamental uncertainty of the future to become more prepared for diverse challenges with adequate lead time" (the time between start and conclusion of a project), often with a 5-year to 25-year horizon (Smith and Saritas 2011). As it addresses something complex, foresight has employed a diversity of methods and is more like a toolkit from which to choose than a standard methodology. Instead of trying to predict *a future*, it moved towards describing and discussing different possible ("might happen"), plausible ("could happen"), probable ("likely to happen") and preferable scenarios (Voros 2001). In this way, decisions could be if not certain, then wiser.

Another method, Horizon scanning, appeared in the 1990s, sometimes seen as a part of Futures Studies (Smith and Saritas 2011). It has been adopted by many programmes across the world, serving several purposes, including identifying new technologies or uses of technology that warrant assessment as well as monitoring implementation of technologies (Carlsson and Jørgensen 1998; Sun and Schoelles 2013; Goodmann 2014). Horizon scanning may also aim to "forecast the health and economic impacts of technologies" and "anticipate potential social, ethical, or legal implications of technologies" (Goodmann 2014). It uses a broad range of data sources about a phenomenon to identify perspectives and trends.

The Gartner hype cycle was first introduced in 1995 by an American firm (Fenn et al. 2013). It is a model predicting the development of technologies in five phases of hype (market promotion) as opposed to the actual value of a technology. It is supposed to help identifying what stage a technology is in and predict which ones are next. As a general and adaptable framework, it has been used to analyse hype in many fields. The hype cycle has been widely criticised as being non-scientific, not properly validated, and thus creating an illusion of predictive capacity where some technologies do not follow its trajectory. At the same time, it may be useful as a crude heuristic as it clearly relates to phases a technology can go through.

Other more scholarly approaches are more grounded in the humanities and social sciences. For instance, the sociology of expectations arose around 2000 (Brown and Michael 2003; Borup et al. 2006). It brought fresh perspectives that influenced TA, particularly TA with a critical-hermeneutic perspective (see below). It drew on theory from science and technology studies (STS), and what had been called the sociology of the future, but also perspectives and methods from history, economics, innovation studies and philosophy of science. It was particularly concerned with promises and expectations in the health and life sciences, and thus relevant to any future-oriented form of HTA. It helped shift the focus in TA from trying to predict the future towards examining the visions of the future as they appear in the present and how they change through time. The sociology of expectations emphasises how visions and expectations have consequences today, how they are used by different agents over time to further their interests (often legitimate but sometimes not), and how failed hype has damaged the reputation of science (Brown and Michael 2003). In this way, it influenced for example vision assessment, which is central to this book. A similar approach is taken, for instance, by the analysis of "sociotechnological imaginaries" (Jasanoff and Kim 2015).

Interestingly, health technology assessment (HTA) as introduced in Sect. 1.3 is not a straightforward subfield of TA, but more a regulatory and commercial prerequisite for any medical innovation that wants to make it through clinical practice. The social view of technological development and its consequences typical of TA seems not to have been shared by the dominant strand of HTA, which focused more on established evidence and relied on EBM (Palm and Hansson 2006). Historically, however, another strand of HTA practice has been developed that, in some respects is more similar to original TA than the dominant form of HTA. It has roots in science and technology studies and applied ethics, seeking to integrate social issues and ethics as well as qualitative and philosophical methods from the humanities with "traditional" HTA methods (Grunwald 2011; Oortwijn and Sampietro-Colom 2022). This leads us to the methodological framework we found most suited to our aims.

2.2 Vision Assessment

Attempts at predicting and pre-emptively controlling technology developments through technology assessment methods face a fundamental problem: the future is uncertain. Lucivero, Swierstra, and Boenink (2011) put this succinctly:

> Such forms of 'upstream' TA by definition have an elusive object: a technology that is only emerging and thus, as yet, exists mainly in the form of visions, promises and expectations. These may pertain to how the technology will look like in the laboratory, how it will perform on the market, how people will use it, profit from it, and how human life will be improved as a result. Unfortunately, experience attests that these expectations at best provide a shaky basis for deliberation and decision-making.

At the same time, there seems to be a relatively broad recognition that we do need to somehow engage with visions of the future and assess emerging technologies before there is abundant evidence (Brey 2012).

Vision assessment was presented as a framework for such engagement in a 2000 book edited by John Grin and Armin Grunwald (2000). Grunwald and others have later developed the related concept of hermeneutic technology assessment (Grunwald et al. 2023). In our search for methodological footing, vision assessment became the theoretical bedrock on which we build in our present work.

Grin (2000, p. 11) defines visions in terms of two features: "they are a mental image of an attainable future shared by a collection of actors; and they guide the actions of and interactions between those actors." While such visions are based on no or little clinical evidence from observations on a realised technology, they are based on *something*: prior scientific knowledge, theoretical assumptions, predictions, arguments, and value judgements. This makes visions and envisioned technologies fruitful for philosophical and ethical analysis, and another form of technology assessment than the form based only on reviews of mostly quantitative, empirical knowledge. The predictions and visions can impact technology development in that they give rise to political and public expectations that in turn result in different funding and investment priorities as well as legislative changes that may not be thought through: "Such expectations can be seen to be fundamentally 'generative', they guide activities, provide structure and legitimation, attract interest and foster investment… In a sense, expectations are both the cause and consequence of material scientific and technological activity" (Borup et al. 2006, pp. 285–86).

Our option is therefore to not only evaluate existing evidence, but also to perform philosophical and ethical analysis of the theoretical foundations on which its visions rest. In the absence of evidence that enables a "traditional" HTA, vision assessment proposes a methodology to explore, unpack, and assess the promises of an emerging field. In contrast with traditional HTA, which can have direct implications for market authorisations and reimbursement of healthcare, assessments of visions and expectations are more of a prospective and reflexive exercise that will feed the discussion among stakeholders. Even though we cannot precisely know what will happen in the future, a decision about implementing something may still be well-supported and wise, or poorly founded and unethical.

Vision assessment is a prime example of a framework that does not aim to predict the future but instead to analyse the contents of a vision as it is communicated in the present. Its methods are philosophical and hermeneutic. The focus is on uncovering and analysing the argumentative validity of knowledge claims and to uncover the societal meanings—often controversial and conflicting—that are attributed to new and emerging technologies in technological visions (e.g., how plausible, useful, relevant, disruptive it is supposed to be). Hermeneutic technology assessment means treating the visions not as facts, but as narratives or stories that may be interpreted and reinterpreted.

A vision assessment may also uncover implicit epistemic and normative assumptions underlying the *visioneering* related to the visions (Trujillo 2014). Visioneering is a term coined by historian Patrick McCray (2013) to describe the way thought leaders are not only imagining visions, but also "engineering" them, that is, actively seeking to realise them through technological and political change. The main authors and engineers of the visions, called *visioneers*, may change the vision along the way, to serve changing realities and purposes. The vision is not only a desired future, but a tool for enacting change in the present.

In (2009) Grunwald outlines three parts that the vision assessment should consist of:

- vision analysis (uncovering and describing the contents of the vision and how is it used in concrete communication),

2.2 Vision Assessment

- vision evaluation (evaluating and judging the contents of the vision), and
- vision management (asking how stakeholders should deal with the vision).

The first part, the vision analysis, is dedicated to accurately reconstructing the vision and its assumptions to set it up for the evaluation itself. The vision is generally articulated around the promise of the delivery of a certain technological solution for a health issue. This promise itself relies on a set of assumptions that are grounded in the properties of the technical device, on the science, on the inscription of the technique into the medical system, and more generally on elements of the society in which it is planned to be used.

The material for vision analysis is where the vision is formulated ("carrier media") (Grunwald 2020). This is to large extent written texts of a wide variety, but may also be, e.g., scientific diagrams, pictures, oral presentations, audio or and video sources and artistic products. The documents of interest can be identified as those containing simultaneously "narratives of the future" and a description of the new technology. Vision analysis means uncovering, identifying and describing the components that underlie the predictions and promises of a vision, notably basic assumptions, underlying premises, conditions and prerequisites, narratives, and arguments.

One central methodological element is to identify the metaphors employed in the vision. Metaphors may reflect implicit assumptions about the technology. The thought here is that metaphors are key concepts that are used to explain the vision where other words fail: "A metaphor is one of the means to produce an utterance of a vision", "allowing communication about an object that does not yet exist in reality and making explanations easier" (Mambrey and Tepper 2000). Some metaphors in medicine are more stable over time than others and may link the vision to a broader philosophical trend. The machine metaphor of the human body is one example. Another methodological element is to uncover different versions of the vision where there may be conflict between different stakeholders in the field about a technology, for example about its promised effectiveness.

In the second part of vision assessment, the vision evaluation, the assumptions or promises that have been unpacked in the vision analysis are rigorously tested. Considering the state of the published literature, the limitations and challenges stated in reviews, and the information provided by experts in the field, the researcher ponders how reasonable these different assumptions are and make explicit the challenges towards the realisation of the promises.

The vision evaluation critiques the described vision in a fair manner. Importantly, this also means evaluating any specific evidence for the technology that may so far have been produced, as well as its underlying assumptions, prerequisites, and conditions. In this way, we aim to assess the epistemological uncertainty pertaining to these technologies and their ethical implications.

In our present work we develop and build upon vision assessment as presented in the literature to suit our more specific aims and purposes concerning organoids. In our development of vision evaluation, we want to assess the overall credibility of the promises made and expectations comprising the vision, the degree to which it is realistic or hyped based on the evidence and uncertainties involved.

First, we will focus on how representative the patient-derived tumour organoids can be expected to be. We here evaluate the biological and scientific underpinnings of the organoid models. When evaluating this part of the vision, *model uncertainty* is one important conceptual tool (Djulbegovic et al. 2011). Model uncertainty arises from the difference between a predictive model and the reality it models (problems with its accurate representation). In the context of organoids, it is about the *biological plausibility,* the scientific and philosophical soundness of the promises related to the models, which we assess.

Second, we cannot only consider scientific and theoretical limitations. Also, technical and practical feasibility are important. How relevant is the technology for a specific clinical application? Could it be realised in practice? We want to assess the plausible clinical utility of these emerging technologies. By *clinical utility* we here mean their feasibility and usefulness in a complex, clinical and social setting. This is wider than, but includes the classical components of HTA: efficacy, effectiveness, and cost-effectiveness. *Efficacy* refers to what can be measured under the circumstances of clinical trials, while *effectiveness* refers to what can be measured in clinical practice. Additionally, we will consider *cost-effectiveness*, although we expected evidence of this to be sparse or non-existent so far.

Finally, an important aspect for us in our unpacking of epistemological uncertainty in the vision evaluation, is the extent to which the envisioned outcomes of organoid research can be assessed through methods that are seen as creating reliable knowledge. This may seem trivial, but it is not. A technology may in theory be very useful, but impossible to show as such. It may also take a very long time before sufficient evidence can be gathered. Can the promised outcomes plausibly be documented or are there types of uncertainty that hamper knowledge production for these outcomes? This includes an important epistemological question about how clinicians can know if a treatment will work in an individual patient, based on information coming from personalised models. If the models are useful, how can we assess and document that they are? We will here begin by considering how far the documentation process has come so far by looking at the published clinically relevant evidence. In this part of the evaluation, uncertainty about the evidence (Djulbegovic et al. 2011) forces us to assess the evidence status of the vision. We will then move on to explore what planned and registered research is in the "translational pipeline" and can be expected to be known in the near future. Finally, we consider obstacles and limitations to knowledge as well as ways forward.

The third part, vision management, is meant to provide advice to stakeholders to guide technology development in the making. Based on the vision evaluation, the vision management will point out ethical and societal issues that need to be addressed or at least acknowledged going forward. Here we consider the responsibility of the vision (Sand 2018). By responsibility we mean how responsible the *promotion* of the vision is, as this is what can be assessed in the present. Does the promotion of the vision go against basic ethical principles such as truth-seeking, honesty, justice and beneficence? Is the potential harm to patients or other stakeholders minimised?

Who will really be able to benefit from the technology? Here we gather any ethical implications of the epistemological issues we have come across in our evaluation. Particularly, we will ask if there is undue hype (overpromising) or downplaying of the potential (underpromising), and injustice in the vision. We ask a question about the future impact: if the vision were to come to be realised, could there be outcomes that make the vision irresponsible?

As a summary, we keep this list of questions, some adapted, and some directly borrowed, from Grunwald (2014) in Table 2.1.

Table 2.1 A checklist for vision assessment of emerging technologies

Description
What is the content of the vision?
Which perceived problem is the vision responding to?
Which concepts and patterns of communication are employed? In which narratives are visions communicated? Which language patterns are employed?

Analysis
What are the epistemic and ontological assumptions underlying the proposed strategy/solution? Which images of humans, social designs, or images of nature are imputed? Which assumptions about the relationship among humans, nature, society, and technology and their future modifications do they contain?
Which assumptions do they contain about the roles, tasks, power, and responsibility of science and technology (do they for example assume technological determinism)?
Where do the visions come from? Who are the authors? With which scientific, ideological, or political backgrounds are they linked? In which networks are they active? Which interests do they pursue?

Challenges
What is the specifically new aspect about the emerging technology being discussed? How does the emerging field differ qualitatively from the known types of technology?
What are the challenges of realising the visions—unarticulated as well as stated by visioneers and practitioners in the field?

Implications
Which spheres of society will be affected by the emerging technology, and which questions does this pose? Are the rights of present-day or future individuals affected? Which questions are relevant for the responsible research and innovation debates and why?
What are the potential social and ethical implications of overestimating the plausibility, amenability and relevance of the visions? What are the potential social and ethical implications of overestimating the plausibility, amenability and relevance of the visions?

2.3 Other Relevant Approaches to Assess and Guide Technology Development

Before turning to how we structured our vision assessment methodology, we will mention some other approaches that also strive towards ethical assessment of emerging technologies.

For instance, Paul Sollie (2007) describes what he calls a "proactive ethical assessment of technology development," emphasising that uncertainty is underappreciated in ethics. Sollie argues that ethics often takes for granted that sufficient information is present to guide moral reasoning. Evaluating emerging technologies challenges this assumption and poses problems for substantive theories of ethics, which seek to make ethical assessments based on factual premises about the world. Instead, it invites procedural theories, which are about "testing the validity of hypothetical norms," and about what is a good procedure of decision-making.

The "techno-ethical scenarios" approach (Boenink et al. 2010) sought to develop a framework for identifying and evaluating possible future "soft" impacts of emerging technologies, for example, on social relations, moral values and human identities as opposed to "hard" impacts, meaning more measurable impacts on health and safety. Analysis of soft impacts addresses how moral values are influenced by technology development and not only vice versa. In the same vein as vision assessment, Lucivero et al. (2011) also state that ethical assessments of promises should begin with an "epistemological analysis of uncertain futures." They further state that analysis of expectation statements frequently rests on three interrelated types of claims. The first type of claim is about the characteristics and functioning of the technology. One should here distinguish between a persuasive discourse, which is often simplified, and the most "honest" discussions in the scientific community pondering limitations and hurdles. In our approach, we will also look for both kinds of statements and use the latter to shed light on the former. The second type is about societal usability and how it will be integrated in current (medical) practice. Claims like these rely on premises about how society works (clinical practice is also socially embedded). This corresponds roughly to our focus on the practical feasibility of using organoids for PM. The third type of claims are about desirability, that is, how the technology will address a social problem or need, and how different audiences will evaluate the morality of the envisioned technology. This corresponds both to our focus on practical feasibility and our vision management part.

Several other frameworks could be mentioned, such as the "anticipatory technology ethics" (Brey 2012) providing both something concrete in terms of methods but also flexibility so that it may be tailored for different purposes. Of note, many approaches, while not being primarily methods for assessing technology, focus on the best way to embed the ethical assessment at the early stage of technology emergence and foster reflection and integration of ethical thinking in the engineering and research process. That would be the case for constructive or participatory technology assessment mentioned above, real-time technology assessment (Guston and Sarewitz 2002), midstream modulation (Fisher et al. 2006) or ethics-by-design (Brey et al. 2021). Integrating ethical reflection all along the development process is a good thing, but the issue of how to best describe and assess the vision remains if we do not want to err on the side of speculative ethics.

2.4 Material and Methods

We now move on to describe more concretely how we went about conducting our vision assessment. Again, we do this in some detail to be transparent and to guide other researchers who might want to follow in our path, or develop it further, in the future (readers who are primarily interested in organoids may move faster through these pages).

We rely on the analysis of four types of written material:

- **Review material**: a material of reviews and other publications—excluding original experimental studies—coming from within the field itself. They contain descriptions of the vision and sometimes also take a reflexive stance with players in the field describing challenges and limitations they are facing.
- **Epistemological hotspots material**: A material of transcribed interviews with central researchers working in teams and studies at the cutting edge of clinical translation. We term these "epistemological hotspots."
- **Clinically relevant published evidence**: Original publications reporting published clinically relevant studies. Here, we not only looked for any recent randomised controlled trials (RCTs), which we expected to be none or very few (Hofmann et al. 2023), but non-RCT clinically relevant evidence.
- **Clinically relevant registered trials**: A material consisting of clinically relevant registered trials that may say something about upcoming evidence ("the evidence pipeline"). This includes both registered RCTs and non-RCTs.

2.4.1 Development and Analysis of Review Material

As mentioned above, the recommended strategy for vision assessment is to start from "carrier media," especially the scientific review, perspective and opinion literature, where scientists and visioneers speak most freely. The aim was not to develop a complete material of all existing references in the world treating organoids in relation to PM, but one that would adequately reflect the vision. Notably, the gathering of the material was done in the context of the HYBRIDA project, where we also focused on organ-on-a-chip (OoC) technology and regenerative medicine.

A search was conducted using the PubMed database, designed to pick up publications at the thematic intersection between organoid models on the one hand, and

PM on the other.[1] The search yielded 1135 hits. Titles, whole abstracts and in some instances parts of, or the whole, text were read for these hits.

Review articles, perspective articles, analysis, opinion articles, editorials as well as journalistic reviews and journalistic articles in English were included (although other types of publications were included, we called this "the review material" for simplicity). References were broadly included that focused on PM as a result from organoid and OoC research. References did not have to have PM as a result from organoids or OoCs as a main topic, but to be included it needed to seem a substantial point in the text relevant for our aims. Original empirical articles were excluded from this material. Secondary material that concerned the ethical and/or epistemological aspects of the vision and were written by scholars outside the organoid/OoC fields were also excluded (but kept as secondary material to refer to in our discussion). A small number of Chinese, French, and German language articles were excluded. In the end, 375 references were selected from the PubMed search.

In addition to these references, we obtained some references through diverse means and "snowballing" throughout the research period[2] (e.g., reference lists and tips from colleagues). This yielded an additional 23 references. In total our searches yielded a material of 398 included references.

224 of these articles were either general reviews about organoids for precision medicine, often with a substantive focus on cancer (53), or specifically about cancer organoids (171). The set of 224 articles is the part of this material that is most relevant for this book. Other articles are specifically on non-cancer disease, on organoids representing specific organ systems, regenerative medicine or on organs-on-chip.

The whole review material was uploaded as PDF files to the analysis software NVivo. In NVivo, the material was analysed by coding it through an extensive set of codes, that is categories of text that pertain to the same question, theme, or point. The codes were divided in two: (1) Codes pertaining to vision analysis, (2) Codes for vision evaluation. The codes were generated both in an inductive fashion from full reading of publications or through coding, and a theory-driven, deductive fashion where we used our prior theoretical understanding to generate code. Codes were

---

[1] In the original search terms, we were including organs-on-chips (OoCs) as well. As noted in the introduction, OoC were then discarded in the analysis. The original search was the following: ("microphysiological systems" OR "microphysiological system" OR "organoid" OR "organoids" OR "organ-on-chip" OR "organs-on-chip" OR "organ-on-a-chip" OR "organs-on-a-chip" OR "organ on chip" OR "organs on chip" OR "organ on a chip" OR "organs on a chip" OR "organs-on-chips" OR "multi-organ-on-a-chip" OR "human on a chip" OR "human-on-a-chip" OR "human-on-chip" OR "human on chip" OR "body-on-chip" OR "body-on-a-chip" OR "tumour-on-chip" OR "tumour on chip" OR "tumour-on-a-chip" OR "tumour-on-chip" OR "metastasis-on-a-chip" OR "metastasis-on-chip" OR "metastasis on chip" OR "cancer-on-a-chip" OR "cancer on a chip" OR "cancer-on-chip" OR "cancer on chip") AND ("personalised medicine" OR "personalised medicine" OR "p-medicine" OR "precision medicine" OR "individualised medicine" OR "individualised medicine" OR "stratified medicine" OR "personalised model" OR "personalised model" OR "personalised models" OR "personalised models" OR "n-of-1" OR "n = 1"). The search was conducted at three different time-points, the last being 24.10.2022.

[2] Until April 2023.

developed and adjusted in a dynamic way between these inductive and deductive approaches.

As the material was too voluminous for full reading of all references, extensive use was made of NVivo's text search functions. The PDF material was searched using various search terms locating pieces of text containing those terms. From the total material, a smaller number of publications was read in full. They consisted primarily of publications that had PM as a main focus, that were generally newer to reflect the state of the art, that had been extensively coded using searches, that were often comprehensive, looking broadly at the field, and that were often written by key figures in the field, preferably for major journals.

2.4.2 Development and Analysis of Epistemological Hotspots Material

The second part of our material was a set of transcribed semi-structured qualitative interviews with centrally positioned researchers working at what we termed "epistemological hotspots". The hotspots are locations where we surmised that the epistemological and ethical questions we are asking are being faced in practice, and where researchers are well-positioned to reflect on them. The hotspot could be:

- An institute/hospital/department where clinical research is located/institutionally supported
- A trial, a series of trials, a past or ongoing translational project
- A research group with a group leader specifically involved as a visioneer in the field

Our search for "epistemological hotspots" began with the analysis of the "review material" and consulting experts in the HYBRIDA group. Based on this, we made a list of "epistemological hotspots" with central members. These experts all were—or have recently been—e.g., principal investigators of a relevant clinical trial, authors of landmark papers, prominent figures in the field with many contributions and editorial activity, etc. The selection procedure was strict in the sense that we contacted only persons with a significant contribution to the field who *also* engaged in clinical translation involving patients, which is not a combination that many can claim.

Interviews are a good way to elaborate insights from the review material: Beyond the written statements in the literature, what do they envision for the future of the field, from concrete prospects to wildest dreams, from positive expectations to grave reservations? We recruited 10 interviewees with whom we spent more than 1 h on average for in-depth exchanges. A semi-structured interview guide was elaborated with two main parts, one corresponding to the vision analysis and one corresponding to vision evaluation parts of the vision assessment. The interviews were then transcribed and analysed.

On top of these formal interviews, we benefited from many other more informal interactions with researchers and clinicians that have occurred during this work.

2.4.3 Development and Analysis of Material of Published Clinically Relevant Evidence

To find and specify the publications we were looking for in our search for the available clinically relevant evidence, we developed the following inclusion and exclusion criteria.

Inclusion:

- Publications providing evidence that related findings in organoid technology to a clinical outcome in patients in a way that we deemed relevant to the management of patients in the clinic. This includes studies that were not only called clinical studies, but also preclinical studies.
- Trials that related findings in the organoids or related technologies not only to clinical outcomes generated in the same trial, but also to published outcome data from other trials.
- Both observational and interventional trials, and also single-case design studies (SCDs).

Exclusion:

- Publications that correlated findings in organoids and related technologies only to, e.g., genetic, histological, or biochemical findings in patients and not clinical outcomes.
- Studies without human subjects.
- Publications focusing on spheroids (which are not fully organoids, as we want to assess the envisioned, more complex models) and explants (tissues that are not self-renewing like PDOs and simply transferred from the body) as outside our scope.
- Publications only focusing on patient-derived xenografts (PDX) models.
- Conference abstracts and preprints.
- Non-peer reviewed publications and cases only described in, i.e., reviews, news and opinion articles.

We could not devise a search string for a search engine that would be both sensitive and specific enough in picking up trials (e.g., search for specific kinds of trials). Neither could we go through all publications on organoids as they would be too many. We therefore used the following three-point strategy to identify publications. Firstly, we performed a text search in all the review material references from 2021 and 2022 using the search "'clinical trials" OR "clinical trial" OR "clinical study" OR "clinical studies"'. We located all the text passages where these terms were used and coded them, assuming that most clinically relevant publications in the field would

2.4 Material and Methods

be mentioned in reviews during the last two years. We then analysed these codes as well as text passages that had been coded earlier in the work as "empirical results" to locate references. This resulted in a list of 94 references. Through further reading of titles, abstracts and in some cases of doubt the full text, 58 references were excluded using the inclusion and exclusion criteria. The result was 36 references. Secondly, to locate references that could not be obtained by consulting our review material, especially newer publications not covered even in recent reviews, we used the same PubMed search string described above that we used to gather the review material, only now to locate original, clinically relevant empirical publications. We assumed that we had located most or all of the relevant older publications from looking at the review material, and we therefore opted to search only for publications registered in the years 2020–2022. This search was last performed 12.12.22 yielding 890 hits. Using the inclusion and exclusion criteria, we focused on locating references that were not already located in the review material, obtaining 15 references. The majority of these were new studies, indicating that our strategy of using the review material to locate older references had been successful. Thirdly, we used snowballing, picking up references in the reference lists and through colleagues that we had not already located throughout the research period until the stop of the writing.[3] This yielded an additional 7 references. In sum, we identified 58 references that we deem to be a comprehensive representation of the evidence status of organoids and related technologies as emerging technologies for the clinic. 55 of these 58 references are about organoids, 3 on organ-on-a-chip technology. None are on regenerative medicine. Cancer was the focus of 47 of the 55 publications. These 47 constituted this part of the material.

The references were analysed using first an Excel-sheet. The publications were all read partially or in full to interrogate the following questions from the material:

- What type of cancer disease and/or organoid did the study consider?
- What type of study design did the publication describe?
- What clinical aim(s) did the study have?
- How many patients/cases were included in the study all in all?
- How many patients who also had an established organoid were treated/considered?
- How many patients were treated?
- How many organoids were developed for comparison?
- How many and which intervention(s) were used?
- What was the main outcome(s) in the organoid (read-out)?
- What were the main outcome measures in the cases/patients?
- What was the follow-up time?
- What was the predictive capacity of the patient-derived model?
- What was the main clinical conclusion?
- What were the main challenges with the study?

From the analysis in the Excel sheet, summaries of the current non-RCT evidence were made and patterns in the material were identified and further analysed. The list

---

[3] Through May 2023.

of publications cannot be seen as fully exhaustive, but we believe that we obtained the large majority of relevant publications through our strategy. The list is available in Appendix A.

2.4.4 Development and Analysis of Material of Registered, Clinically Relevant Trials

To get a sense of the evidence that may be upcoming, in what we may call the "translational pipeline", we developed a material of trials registered on international trial databases containing the word "organoids". Since the purpose is to conduct a vision assessment, we find a mapping of ongoing and planned trials relevant as indicative of what RCT and non-RCT evidence can be *expected* concerning the use of organoids as clinical models to inform decision-making.

A search using the term "organoid" in the trial description on the NIH international clinical trials (CT) database, clinicaltrials.gov, gave 157 results. The same search on Euroregister gave 13 results and the WHO clinical trials database gave 0 results. One of the trials registered on Euroregister was also registered on the CT database and was counted as only one. This gave 179 trial registrations in total, all on organoids.

Among the 179 trials, 8 were listed as withdrawn (e.g., due to lack of funding) or terminated (e.g., due to low accrual rate or feasibility issues). 5 trial descriptions contained the word "organoid" but did not involve organoids in the experimental or clinical study design and were therefore excluded. Inclusion and exclusion criteria were calibrated with the development of the published clinically relevant material. At the same time, the level of detail in trial registrations are often different from that of published abstracts. We included registered trials that we perceived as aiming to use organoids to guide management of patients, but some of these trials may not use organoids in the direct management of patients. This means that the inclusion here may be somewhat broader than in the published clinically relevant evidence material.

57 trials studied organoids only for research purposes, such as preclinical disease modelling or biobanking. As we are interested in evaluating existing and forthcoming evidence for organoids as *clinical models*, i.e., models that are evaluated against a clinical endpoint and used to inform clinical decision-making, these 57 trials were not analysed in detail.

The remaining 109 trials described studies of organoids as clinical models. They were categorised according to different types of trials. As seen in Figs. 2.1 and 2.2, only 16 trials[4] were listed as either completed (5) or beyond the estimated study completion dates with a status as "unknown" (11). For all completed trials, we searched for published findings in online databases and emailed the contact person of the trial for more information. 93 trials were ongoing or planned. They were also analysed although we cannot evaluate their outcomes yet. Since the aim is to look at epistemological uncertainty related in a broad sense and conduct an amended HTA

---

[4] As of February 2023.

2.4 Material and Methods

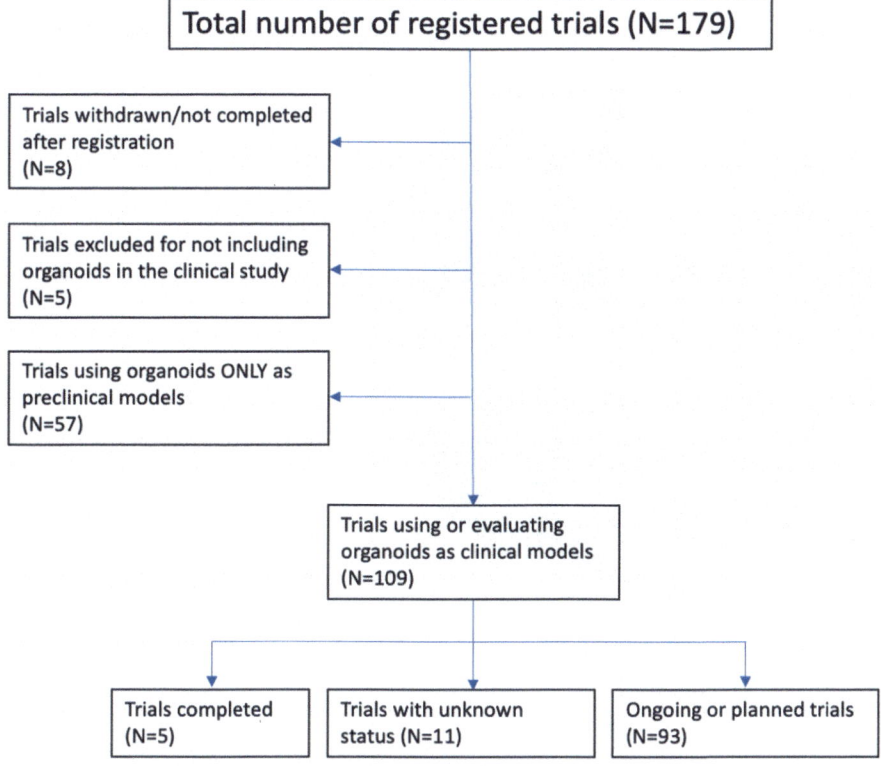

Fig. 2.1 The inclusion and exclusion of registered trials in the material

containing an evaluation of the visions of using organoids in clinical contexts, the characteristics of currently registered trials are interesting, nonetheless.

2.4.5 *Further Development of Vision Analysis and Vision Evaluation*

In developing our vision analysis (Chapter 3), we relied mostly on the review material, but supplemented with the other materials (e.g., interviews). We identified subvisions and their main overarching promises and then identified main expectations underlying each promise, and for each of these expectations more explicit, specific promises. For these specific promises in turn, we identified underlying theoretical assumptions, as well as scientific and technical prerequisites that would be what we could then evaluate in the vision evaluation. We look for key metaphors and concepts, points of conflict where different versions of the vision exist and seek to delineate the history of the vision.

Disease studied in registered trials

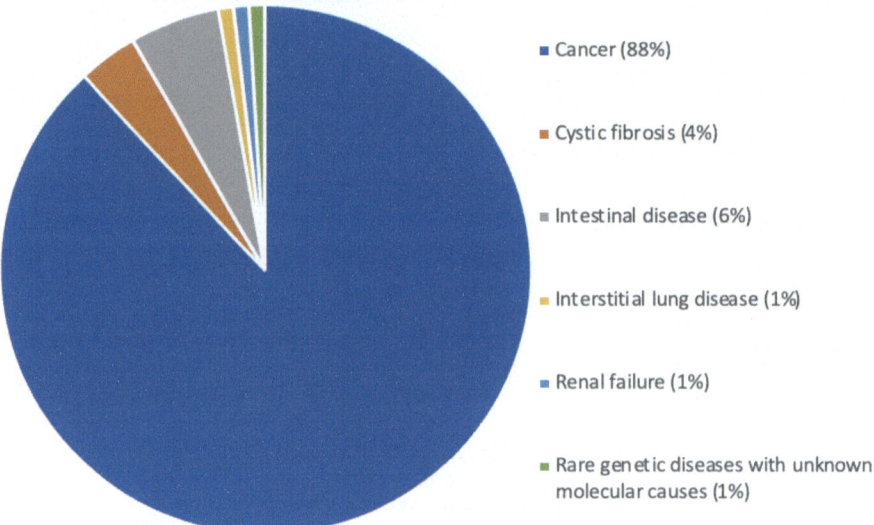

Fig. 2.2 Disease studied in registered trials. In this book we focus on cancer, but the figure is included to show the proportion of cancer research in comparison to other diseases in the organoid field

In developing our vision evaluation method (Chapter 4), we especially searched for and coded passages of text in our review material that contained stated limitations and challenges, i.e., where the authors speak explicitly about the challenges and limitations facing their field. We propose this as a fruitful avenue of inquiry for anyone evaluating a field in which they are not themselves experts. Specifically, the review material that was published in 2020 or later was queried using the search terms "limitation," "challenge," and "hurdle" (which also pick up "limitations," "challenges," and "hurdles" in plural). The text passages with these terms were read and coded in order to systematise the statements on the challenges facing the field. They were then analysed, identifying the various types of challenges and limitations that the authors wrote about. The resulting overview of the limitations and challenges that the authors themselves state provided us with a basis for further critique of the vision based on our own theoretical understanding and perspectives based on the rest of the material (published evidence, translational pipeline, epistemological hotspots). It also enables us to compare the bolder statements about what organoids might achieve in the future, with more reserved statements that may be contradictory.

References

Boenink M, Swierstra T, Stemerding D (2010) Anticipating the interaction between technology and morality: a scenario study of experimenting with humans in bionanotechnology. Stud Ethics Law Technol 4(2)

Borup M, Brown N, Konrad K, Van Lente H (2006) The sociology of expectations in science and technology. Technol Anal Strat Manag 18(3/4):285–298

Brey P (2012) Anticipatory ethics for emerging technologies. NanoEthics 6:1–13

Brey P, Dainow B, Erden Y et al (2021) SIENNA D6.3: methods for translating ethical analysis into instruments for the ethical development and deployment of emerging technologies. Zenodo. https://doi.org/10.5281/zenodo.5541539

Brown N, Michael M (2003) A sociology of expectations: retrospecting prospects and prospecting retrospects. Technol Anal Strat Manag 15:3–18

Carlsson P, Jørgensen T (1998) Scanning the horizon for emerging health technologies: conclusions from a European Workshop. Int J Technol Assess Health Care 14(4):695–704

Djulbegovic B, Hozo I, Greenland S (2011) Uncertainty in clinical medicine. In: Gifford F (ed) Philosophy of medicine, vol 16. Elsevier, Amsterdam

Fenn J, Raskino M, Burton B (2013) Understanding Gartner's hype cycles. Gartner. Foundation document, available online: https://www.gartner.com/en/documents/3887767

Fisher E, Mahajan RL, Midstream MC (2006) Modulation of technology: governance from within. Bull Sci Technol Soc 26(6):485–496

Goodman C (2014) HTA 101: National Information Center on Health Services Research and Health Care Technology (NICHSR). Available from: https://www.nlm.nih.gov/nichsr/hta101/

Grin J (2000) Introduction: vision assessment to support shaping 21th century society? Technology assessment as a tool for political judgement. In: Grin J, Grunwald A (eds) Vision assessment: shaping technology in 21st century society towards a repertoire for technology assessment. Springer, Berlin, pp 9–30

Grin J & Grunwald A. 2000. Vision Assessment: Shaping Technology in 21st Century Society, Towards a Repertoire for Technology Assessment. Berlin: Springer.

Grunwald A (2009) Technology assessment: concepts and methods. In: Meijers A (ed) Philosophy of technology and engineering sciences—handbook of the philosophy of science, vol 9. Elsevier, Amsterdam, pp 1103–1146

Grunwald A (2011) Responsible innovation: bringing together technology assessment, applied ethics, and STS research. Enterp Work Innov Stud 7:9–31

Grunwald A (2014) The hermeneutic side of responsible research and innovation. J Responsible Innov 1(3):274–291

Grunwald A (2020) The objects of technology assessment. Hermeneutic extension of consequentialist reasoning. J Responsible Innov 7(1):96–112

Grunwald A, Nordmann A, Sand M (2023) Hermeneutics, history, and technology: the call of the future. Routledge

Guston DH, Sarewitz D (2002) Real-time technology assessment. Technol Soc 24:93–109

Hofmann B, Zinöcker S, Holm S, Lewis J, Kavouras P (2023) Organoids in the clinic: a systematic review of outcomes. Cells Tissues Organs 212(6):499–511. https://doi.org/10.1159/000527237

Jasanoff S, Kim S-H (eds) (2015) Dreamscapes of modernity. Sociotechnological imaginaries and the fabrication of power. Chicago University Press

Lucivero F, Swierstra T, Boenink M (2011) Assessing expectations: towards a toolbox for an ethics of emerging technologies. NanoEthics 5(2):129–141

Mambrey P, Tepper A (2000) Technology assessment as metaphor assessment. Visions guiding the development of information and communications. In: Grin J, Grunwald A (eds) Vision assessment: shaping technology in the 21st century. Springer, Berlin, pp 33–51

McCray W (2013) The visioneers: how a group of elite scientists pursued space colonies, nanotechnologies, and a limitless future. Princeton University Press, Princeton

Nazarko L (2017) Future-oriented technology assessment. Procedia Eng 182:504–509

Oortwijn W, Sampietro-Colom L (2022) The VALIDATE handbook—an approach on the integration of values in doing assessments of health technologies. Radboud University Press

Palm E, Hansson SO (2006) The case for ethical technology assessment. Technol Forecast Soc Chang 73:543–558

Sand M (2018) Futures, visions and responsibility—an ethics of innovation. Springer, Wiesbaden

Smith JE, Saritas O (2011) Science and technology foresight baker's dozen: a pocket primer of comparative and combined foresight methods. Foresight 13(2):79–96

Sollie P (2007) Ethics, technology development and uncertainty: an outline for any future ethics of technology. J Inf Commun Ethics Soc 5(4):293–306

Sun F, Schoelles K (2013) A systematic review of methods for health care technology horizon scanning. AHRQ Publication No. 13-EHC104-EF. Quality AfHRa.

Trujillo L (2014) Visioneering and the role of active engagement and assessment. NanoEthics 8:201–206

Urueña S (2021) Responsibility through anticipation? The 'future talk' and the quest for plausibility in the governance of emerging technologies. NanoEthics 15:271–302

Vogt H, Gaillard M, Green, S (2022) HYBRIDA D2.3 Adaptation of health technology assessment (HTA) to evaluate organoids and organ-on-a-chip as emerging technologies in the clinic. https://hybrida-project.eu/deliverables/

Voros J (2001) A primer on futures studies, foresight and the use of scenarios. Prospect Foresight Bull 6

Open Access This chapter is licensed under the terms of the Creative Commons Attribution 4.0 International License (http://creativecommons.org/licenses/by/4.0/), which permits use, sharing, adaptation, distribution and reproduction in any medium or format, as long as you give appropriate credit to the original author(s) and the source, provide a link to the Creative Commons license and indicate if changes were made.

The images or other third party material in this chapter are included in the chapter's Creative Commons license, unless indicated otherwise in a credit line to the material. If material is not included in the chapter's Creative Commons license and your intended use is not permitted by statutory regulation or exceeds the permitted use, you will need to obtain permission directly from the copyright holder.

Chapter 3
Vision Analysis of Patient-Derived Organoids for Personalised Medicine

In the following two chapters, we present our assessment of the vision of using patient-derived tumour organoids (PDTOs) for precision cancer medicine or precision oncology (PO). It is a main part of the more general vision of using organoids for precision medicine. In the present chapter, we describe the vision in the vision analysis described methodologically above. We will first present what we perceive as the overarching promise of this vision, and then divide it into three main expectations on which that promise rests. We then further divide these expectations into more specific promises and underlying assumptions, prerequisites and conditions that are amenable for more specific critique (see Fig. 3.1).

3.1 The Overarching Promise: Improved Predictions and Clinical Utility Through Personalisation

To understand the vision we are examining here, let us first explain how it enters the larger vision of personalised medicine (PM). Personalised medicine in a broad sense goes back to the beginning of the medical profession (Hofmann 2003). However, in the late 1990s the concept of "personalised medicine" was picked up by biomedicine in the context of the Human Genome Project and given a specific biomedical meaning (Tutton 2012). The promise was that deep knowledge about the genes and other biological characteristics of the individual case would allow the identification of the right treatment for a given person. Personalised medicine is often contrasted to the purported "one-size fits-all approach" of current evidence-based medicine (EBM), in which everybody is often treated according to the same guidelines (Beckman & Lew 2016).

The vision of precision medicine through organoids may be seen as the convergence of the "mainstream vision" of creating PM through genomics and systems biology (see e.g., Hood and Flores 2012), with stem cell technology developing into

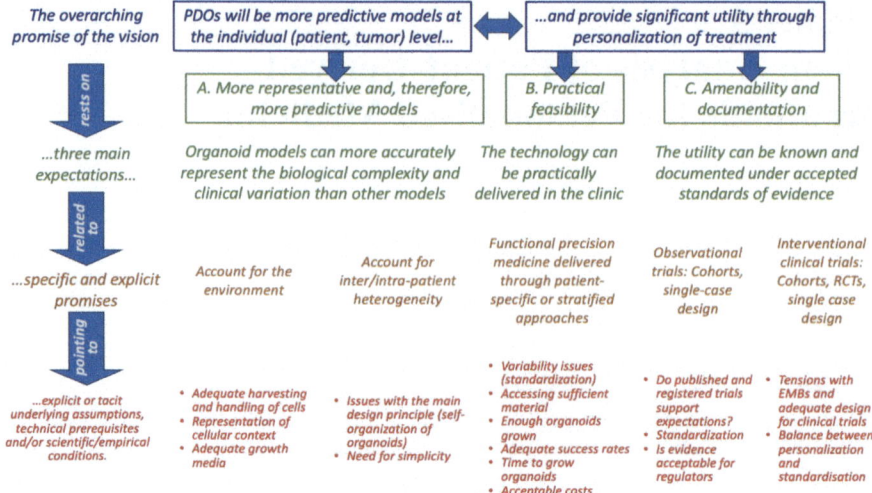

Fig. 3.1 This figure shows the overall structure of our vision analysis, which also forms the basis of our vision evaluation. There are four layers: First, an overarching promise, which is what we aim to assess in the end. The overarching promise rests on three main expectations, which each in turn are associated with a set of more specific and explicit promises. Below these in turn we find various assumptions, prerequisites, and conditions that are often tacit, but which must be fulfilled for the vision to be successful. The list of these underlying prerequisites, assumptions, and conditions is not exhaustive

organoid technology around 2010. Building on decades of cell culture, organoid research gained momentum from 2010 (Simian and Bissel 2017, HYBRIDA D1.3, D2.1). As a landmark in this development, intestinal organoids were established first in mice (Sato et al. 2009) and then in humans (Jung et al. 2011). Shortly thereafter, the vision of using organoids to enable PM, specifically precision oncology, appeared (Campbell et al. 2011). As illustrated by its perhaps most central "visioneer," Hans Clevers, in (2013):

> With the wild-type and colon cancer organoids, we can potentially predict patient outcome and response to drugs. In the future, we hope to rapidly build large, living biobanks for other cancers, too.

Although the means of personalisation are quite different from the original gene-based form of PM, the PDTO for PO vision is another chapter in the mainstream PM idea of using biological information to tailor treatment against established disease. As one of our interviewees from the "epistemological hotspots" material states: "We are talking about putting the right drug in the right patient" (senior researcher, academic). In the vision concerning cancer this is often expressed as a promise of accounting for *inter-patient heterogeneity* or *inter-tumour heterogeneity*:

> One drug does not fit all; differences in drug response due to intra-individual genomic diversity, intratumor and intertumor heterogeneity led to the failure of 'one-size-fits-all' therapies. Therefore, selecting the right drug for each individual should be fit [sic]. To

achieve this goal, reliable cancer models such as PDOs that can mimic the complexities of the original tumor are needed. (Hu et al. 2021)

In keeping with this general theme of PM, we thus find that, the overarching promise in the "PDTO for PO vision" is that these models can *improve predictions of what the patient with cancer disease needs and provide significant clinical utility through personalization of treatment.*

We find that this overarching promise rests on three main expectations.

Main expectation A: PDTOs can yield models that more accurately capture the true complexity of a patient's specific tumour as well as variation between individuals (inter-tumour heterogeneity), and that they therefore will give better predictions for the individual.

Main expectation B: That implementation of the models in the technically and socially complex clinic will be practically feasible.

Main expectation C: That the expected superior predictive capacities and usefulness of the models can also be amenable to scientific documentation through acceptable methods and evidence standards. In other words, the models are expected to create less (epistemological) uncertainty about what individuals need.

Before we further describe these expectations, we will briefly provide some details of the main promise.

3.1.1 Different Stem Cells, Different Aspects of the Promise

The vision of PM based on organoids is based on stem cell technology. However, there are different types of stem cells with different potentials and requirements for development, which complicates the narrative.

3.1.1.1 Induced Pluripotent Stem Cells (iPSCs)

iPSCs have gained widespread popular attention due to the spectacular scientific breakthroughs that have enabled them. They are generated by reprogramming adult cells into a state that resembles that of embryos. This provides great advantages. Organoids based on iPSCs could be generated from any tissue, theoretically in great numbers, contain many different types of cells, and grow into more complex models reflecting more complex biology. In our assessment, however, they only play a limited role where iPSCs can be generated from a person and then gene-edited with mutations so that they resemble the patient's cancer and grown into patient-specific organoids (Hepburn et al. 2020). iPSCs can also be used to develop tumouroids resembling organs from which it is hard to harvest adult stem cells, notably the brain. iPSCs can also be used to grow healthy tissues from a patient, which can then be used to test for patient-specific side-effects of specific treatments, in parallel to testing of the same treatment on organoids representing disease states.

The other form of pluripotent stem cells, embryonic stem cells, which would be harvested from embryos, come with ethical concerns, and form little or no part of the vision as it appears in our material.

3.1.1.2 Adult Stem Cells (ASCs) and Cancer Stem Cells (CSCs)

It is primarily the potential of cancer stem cells (CSCs), that underlies the PDTO for PO vision. They can be seen as a subgroup of adult stem cells.

Adult stem cells (ASCs) can be harvested from most tissues. Unlike iPSCs, they can form only cell types of the tissues from which they are derived. They thus cannot mimic any cell or tissue. Neither do they form all cell types of a tissue. At the same time, they do not need to be made in the lab or the "reprogramming" to start reflecting disease in a person's body. They are relatively easy to develop. They retain epigenetic memory (molecular markings that affect the way protein is transcribed from DNA and perform functions in the body) and reflect the state of mature tissues.

Cancer cells have stem cell-like properties and may therefore be cultured into tumour organoids. They model tumours, not organs. They retain the genetic and histological characteristics of the original tumour and recapitulate the genetic variation seen among cancer patients in the clinic.

We have…

- pluripotent stem cells (PSCs), where we find embryonic stem cells (ESCs) and induced pluripotent stem cells (iPSCs) and…
- adult stem cells (ASC). Under ASCs we find the distinct subgroup of cancer stem cells (CSCs).

3.1.2 A Promise of Precision Oncology

As described in the previous chapter, a large part of our total material is focused on cancer. Reading the titles of our review material, and counting which diseases they mention specifically, we can get an impression of what cancers feature most prominently in the vision (see Fig. 3.2).

From the beginning there is a strong focus on gastrointestinal, prostate and pancreatic cancer in the vision (Gao et al. 2014; Ohta and Sato 2014). Gastrointestinal cancers dominate (this includes gastric, esophageal and colorectal), and, of these, the majority of publications are on colorectal cancer. This may be due to its high prevalence or, more probably, for other reasons related to the history and nature of gastrointestinal organoids (Clevers 2016). However, pancreatic cancer also stands out as a paradigmatic disease, and prostate and ovarian cancers are also common diseases in the visionary literature.

There may be several reasons why certain cancers are prominent in the vision. One of our interviewees from the "epistemological hotspots" points to cancers that have not benefitted from previous "genomic" precision medicine:

3.1 The Overarching Promise: Improved Predictions and Clinical Utility ...

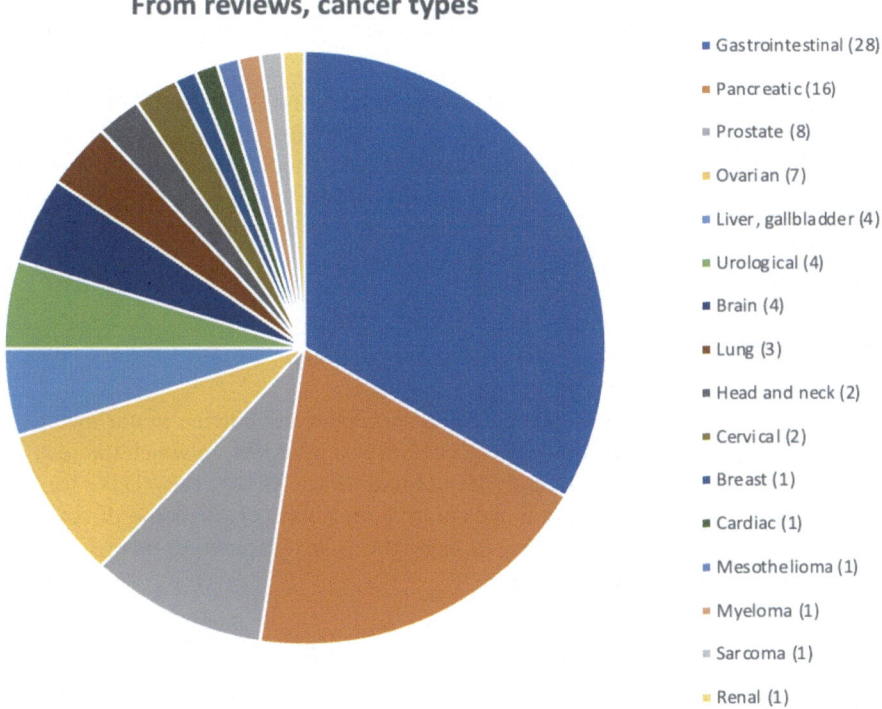

Fig. 3.2 The distribution of different specific cancer types (by organ system) as specified in the titles in our review material. Gastrointestinal cancer here includes gastric, esophageal, gastroesophageal, colorectal, rectal. Liver and gallbladder system includes cholangiocarcinoma and gallbladder cancer. The category "Brain" includes glioblastoma and neuroblastoma in children

> The idea for functional precision medicine was really to go first for cancers that didn't benefit from these breakthroughs. For example, in the case of digestive cancers, if we focus on colorectal and pancreatic cancer, there have been no major improvements since chemotherapy. (senior researcher, academic)

According to another interviewee, the focus is on cancer for which there were few recent therapeutic advances and that has poor prognosis.

> What I see would be those cancer types where we have nothing that can help the patients. Pancreatic cancer has a really bad prognosis—people live a median 6 to 9 months and we really have no drug for them. So, let's see if organoids could help. Another cancer that I can think of is brain cancer, glioblastoma, after first surgery in six or nine months' time they will definitely recur, by the time they recur literally there's nothing you can do, there's no option. (senior researcher, academic)

We will go back to the distribution of cancer types when discussing clinical evidence and evidence assessment.

3.1.2.1 Individualised Drug Screening Versus Personalised Radiotherapy

When visioneers of organoid PM talk about "personalised treatment" they tacitly most often talk about drugs. This is called "individualised drug screening". However, the vision of using organoids to predict the effects of radiotherapy also features quite visibly (personalised radiotherapy) (Wang et al. 2022). This idea is called "precision radiation oncology". The idea here is to radiate the organoid and observe both what is taken to predict positive effects, but also side-effects.

3.1.2.2 Preventive Versus Curative Cancer Medicine

There are no titles referring to preventive or preventative medicine in our material. This stands in stark contrast to the more general vision of PM, in which the idea of prevention features very prominently (Hood and Flores 2012). The PDTO for PO vision is thus mostly a promise about curative medicine, or medicine directed at established disease, which also reduces its boldness as compared to other forms of PM, which aim at prevention.

3.1.3 Different Degrees of Boldness in the Vision

The vision of developing patient-derived organoids for personalised medicine comes in various degrees of boldness when it comes to how much this is presupposed to contribute to medicine and society. For the sake of our discussion, we distinguish between "revolutionary visions" and "modest visions". In the first kind of vision, the visioneers use revolutionary language. For example, that "stem-cell-derived organoid technology (...) is poised to revolutionize anticancer treatment" (Vincan et al. 2018), or that:

> The combination of precision medicine using genomic testing and pharmacological profiles based on high-throughput drug sensitivity testing using patient-derived organoids is expected to revolutionise pancreatic cancer treatment. (Iwasaki et al., 2021)

In the modest versions of the vision, visioneers emphasise models that are "good enough", that fill some specific purpose while not revolutionising patient treatment.

> This technology provides a promising model system to facilitate translational research and may have a role in clinical decision making. (Veninga and Voest 2021)

3.2 Main Expectation A: Improved Prediction Through More Representative Models

We now further describe the three main expectations that underpin the overarching promise.

The first main expectation is that PDTOs can yield models that more accurately capture the true complexity of a patient's specific tumour as well as variation between individuals, and that they therefore will give better predictions for the individual. In the cancer context, the ambition of accounting for biological complexity, is partly expressed as the promise of also accounting for *intra-tumour heterogeneity* and solving "*the challenge of treating many cancers in one*" (Pfohl et al. 2021). This refers to the fact that tumours consist of different types of malignant cells, that may respond differently to treatment (Sasaki and Clevers 2018). Accounting for biological complexity and variation among patients is envisioned to improve predictions substantially.

> Overall, a key promise of organoid technology in clinical applications has been its ability to predict patient outcome—specifically, that drugs with antitumor activity in PDOs would have an analogous effect when treating the donor patients. (Bose et al. 2021, p. 1013)

Here, we must first hasten to declare, that we do not mean to say that players in the field think that predictive power is *always* or *necessarily* linked to more accurate representation, or that they do not understand that models can be useful while not representing a human disease in full (Germain 2014). But we do perceive that, when high hopes are attached to these models, it is because they are expected to be more accurate in representing the true biological complexity of tumours than simpler cell culture models, relying on genetic information alone, or computational models that are not themselves biological entities. Organoids are admittedly simpler than animal models, such as mouse models xenografted with human tumour cells (patient-derived xenograft models, PDX), but they are still hoped to better account for the most important features determining how specific patients respond to treatment. When tumour organoids are promised as more predictive than PDXs, it is because they are assumed to represent more accurately *human biology*. Importantly, organoids are at the same time promised as more representative of patient variation than existing models.

Illustrating this expectation of being representative, organoids are sometimes described through terms such as "*disease in a dish*" (Singh et al. 2021). This is also expressed in the literature's emphasis on increased *physiological relevance* of organoids, compared to existing translational models:

> The common failures to translate promising preclinical drug candidates into clinical success and the endeavors to realize precision medicine highlight the urgent call for a model system with easy accessibility, high physiological relevance and ability to incorporate patient-specific disease background. Organoids derived from stem cells have proven to be such a system.... (Sun and Ding 2017)

To underscore the potential of organoids in representing biology, visioneers frequently contrast PDOs to the strategy of only using the genome (DNA) of a person/disease as a predictive "model" for PM. Generally, in the PM field, it has been increasingly realised that genomic information alone is insufficient to predict what patients need (Chakravarti 2011). Reflecting this sentiment from the organoid perspective, Clara-Trujillo and colleagues (2020) state that, "Personalized cancer therapy has historically focused on profiling tumour DNA, RNA, or protein as molecular biomarkers to predict patient response. However, these methods have not been able to predict therapeutic response". The organoid solution is then presented: "Since PDOs recapitulate more features of human disease than genetics alone, PDO-guided therapeutic decision-making may outperform genome-guided therapies clinically in the future" (Bose et al. 2021, p. 1012–1013).

In the history of PM, there is another field that has previously proposed to account for the complexity of biology and environmental context to build superior predictive models for PM: systems biology and systems medicine (Hood and Flores 2012). Systems biology is characterised by extensive use of mathematical and computational models based on "omics" and other big data to decipher biological systems and thus make predictions. However, this field too is facing challenges (Green and Vogt 2016). Enter organoids, which may be seen as another of biomedicine's promises, coming after systems biology, of accounting for biological complexity. These models offer a way of bypassing the difficult step of generating computational—often called in silico models after the silicone used in computers—models from big data, by using a biological system.

In the vision of PM through organoids, these 3D systems are also often presented as superior to earlier, simpler biological in vitro models based on stem cells as well as immortalised cell lines. As an interviewee says:

> Organoids do a much better job at maintaining the genetic profile and phenotype of an original tumor than if you put them on plastic or two-dimensional cultures, where they start to shift immediately. We really believe that, as long as they (organoids) are handled in the right way, they really can quite accurately represent the actual patients' tumor... So that makes it a really powerful tool for clinical diagnostics, for predicting what therapy might be best for a given patient. (senior researcher, academic)

To fulfil the first main expectation, organoids need to represent a number of aspects of biology, and this is associated with several more specific promises and assumptions, to which we now turn.

3.2.1 The Assumption of Genetic Representation

First in the vision, comes the assumption that organoids can accurately represent the genome of each individual case. As stated by Clevers: "Organoids are a very good genetic representation of the tumour" (Clevers 2015). What this means is that the genetic variation (DNA) among patient-derived organoids continues to be representative of the genetic variation among patient cases as the organoid is grown. So, it will be an accurate genetic model.

3.2.2 The Microenvironment—A Promise of Context and Holism

However, the central promise in the organoid vision is to account for the environment (see e.g. Xu et al. 2021, Zanoni et al. 2020). Biology is often said to be about gene-environment interaction, the interplay of "nature" and "nurture". The organoid field promises not only to be able to account for the genome, that is, "nature", but of "nurture", the context of the genome. As Gilazieva et al. (2020) state, "The main advantage of 3D models is the ability to reproduce more natural conditions of cell growth and interaction". Organoids allow a replication of the immediate molecular and cellular surroundings of cells, and, in the case of cancer, the tumour micro-environment (TME).

> The response to therapy, clinical outcomes, and tumor behavior such as metastases, tumor progression, carcinogenesis can be significantly affected by the heterogeneous tumor microenvironment (TME) and interpersonal differences. Therefore, using native tumor microenvironment mimicking models is necessary to improving personalized cancer therapy. Both in vitro 2D cell culture and in vivo animal models poorly recapitulate the heterogeneous tumor (immune) microenvironments of native tumors. The development of 3D culture models, native tumor microenvironment mimicking models, made it possible to evaluate the chemoresistance of tumor tissue and the functionality of drugs in the presence of cell-extracellular matrix and cell-cell interactions in a 3D construction. (Hu et al. 2021)

We find that, in the vision of cancer organoids for personalised medicine, it is this idea of accounting for the TME that is most clearly linked to the idea of "holism". As already alluded to, systems biology and systems medicine brought a promise of holism against the previously reductionist (as in "gene-centric") concept of biomedical personalised medicine: To account "holistically" for the full complexity of human biology in computational models (Vogt et al. 2016). Here, this idea of holism reappears in a slightly different form, as a promise of providing a "holistic" environment for genes and cells.

> Historically, 2D culture has been used to better understand cellular mechanisms, organ development and function, and disease modelling, amongst many other topics. However, a more holistic approach to modelling has been taken by considering the cell microenvironment and the ECM components that tissue is composed of. (Mazzocchi et al. 2019)[1]

---

[1] ECM means extracellular matrix, a network of proteins and other molecules that support cells and give structure to tissues.

3.2.3 The Assumption of Self-organisation as a Design Principle

How are the patient-derived models that can achieve superior representation promised to be built? Here, the concept of *self-organisation* is paramount.

> The term 'organoid' literally means organ-like, reflecting the ability of organoid culture conditions to prompt cells to self-organize into structures mimicking the architecture of the organ from which they were derived. (Baker et al. 2016)

The idea here is that stem cells, given the right environment, have the capacity to organise or "complexify" themselves into tumour-like forms. To let nature build the models itself through its creative powers, so to speak. To engineer living organisms, the thinking goes, is too complex for humans (yet), but we may let stem cells do it for us. A key premise in the vision of organoids is thus a form of at least partial *self-design*, to "self-organise according to the same intrinsic organising principles as in the organ itself" (Günther et al. 2022).

3.3 Main Expectation B: Practical Feasibility

The second main expectation behind the overarching promise is that the use of more predictive PDOs can feasibly be delivered to patients in clinical, oncological practice.

3.3.1 Functional Precision Medicine, Functional Biomarkers, and Molecular Agnosticism

The concepts of "functional precision medicine" and "functional biomarkers" are important in this regard. It is via functional precision medicine, "that organoids will have clinical applications in real life" (Foo et al. 2022, p. 4). The concept, first introduced by Letai in (2017), is associated with an idea of "moving beyond pure genomics". Here one gives treatments to patients based on observable "functional" responses that appear in the organoid *after* treatment on that model instead of PM based on "static" (non-functional) information about the genome or other molecular biomarkers that are available before treatment:

> Functional precision medicine is a strategy whereby live tumor cells from affected individuals are directly perturbed with drugs to provide immediately translatable, personalized information to guide therapy. (…) Precision oncology has traditionally used static features of the tumor to dictate which therapies should be used. Static features can include expression of key targets or genomic analysis of mutations to identify therapeutically targetable 'drivers'. (Letai et al. 2022, p. 26)

Mutations and other molecular biomarkers are here portrayed as not providing enough information on what to give the patient (Riedesser et al. 2022). "Functional" means that one metaphorically moves from "snapshots" to dynamic "moving pictures" as a basis for decision-making in PM.

> The concept of functional precision medicine, or functional biomarkers, is attracting attention as an alternative way to predict drug efficacy. Rather than exclusively relying on a snapshot of information, functional precision medicine will offer highly actionable and functional information from assessment of the characteristics of viable primary tumor cells. (Kondo & Inoue 2019)

Instead of using DNA or some other molecular biomarker to predict the outcome statistically, one may observe how a treatment actually works in the model and extrapolate that to the patient. This is also called "personalised functional drug testing", "functional drug screening" or "personalised functional medicine assay" (Napoli et al. 2022).

Another related concept is *"molecular-agnostic predictions"* (Betge and Jackstadt 2022). This captures the idea that one can predict what a person needs without genomic or molecular biomarkers.

3.3.2 The Clinical Setup: How Personalised Medicine is Expected to Be Delivered

The biomedical idea of personalised medicine was, historically, strongly tied to the idea of *stratified medicine.* This means grouping people in smaller "strata" that share the same or similar biological characteristics, doing research (e.g. clinical trials) on those strata to gain knowledge about what works for that group, and finally treating future individuals appearing in this group according to this essentially population-based knowledge. However, there is also a more radical version of the vision that seeks to "truly" treat the individual based on his/her biological characteristics held up against biological mechanistic knowledge, and, also using single-case design studies where n = 1 and the person is his/her own control (Tonelli and Shirts 2017; Nikles et al. 2021; Vogt 2025, in press).

In keeping with this historical distinction, we also find two main approaches for use of organoid models in PM. The first we call *the patient-specific approach* with the *"avatar"* concept as a key metaphor (see Fig. 3.3), and the second is what we will call *stratified approaches* (see Fig. 3.4). The main distinction between the two is that in the patient-specific approach one tests a treatment in an organoid from *the same patient* that one wants to treat, while in the stratified approach one tests the treatment on an organoid based on cells from a patient who is *similar*. Combined approaches are also promised.

Patient-derived organoids are often called "personalised disease models" or "personalised models" (see, e.g., Veninga and Voest 2021; Mazzocchi et al. 2019). This concept is used both in the context of the patient-specific and stratified approaches.

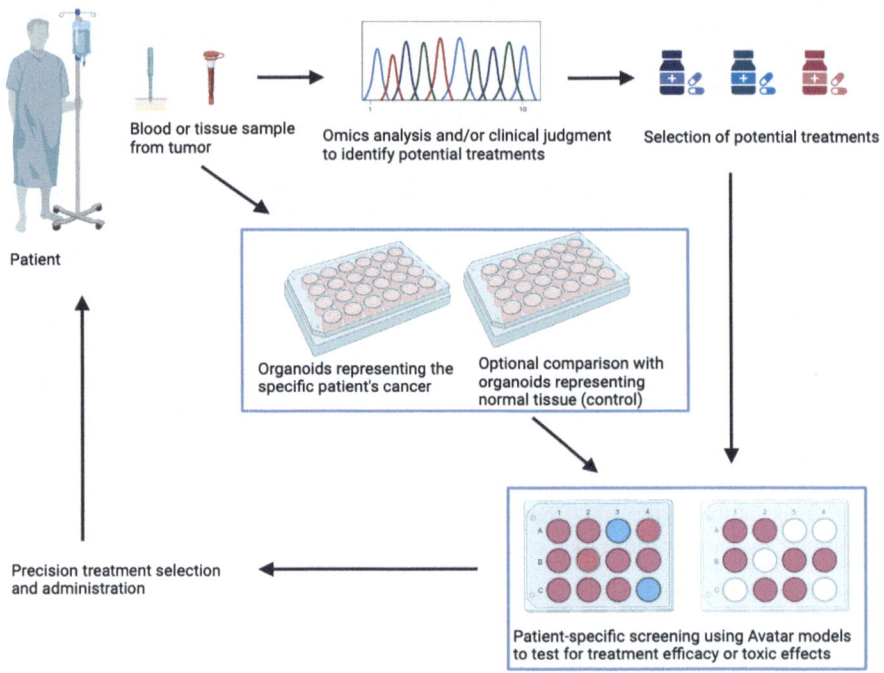

Fig. 3.3 Patient-specific screening using so-called "Avatar models" to test for efficacy and toxic effects of cancer treatments. The figure illustrates the basic principle behind the use of organoids for patient-specific screening, based on our analysis of scientific literature and diagrams within these. Organoids based on tumour samples from specific patients are developed as an assay for testing different candidate treatments, e.g., targeted treatments that match the mutations in the patient's tumour. Healthy cells can be used as controls, making the treatment screening similar to a bacterial infection test, where a good test result is to identify treatments that kill what causes the disease, but not the healthy cells. The figure was created by the authors using BioRender

Our vision evaluation below will therefore be about the ability of these models to predict *as functional biomarkers*, the extent to which this may be coupled to useful interventions in practice, and also documented as such.

3.3.2.1 The Patient-Specific Approach—The Avatar and the Antibiogram

The patient-specific approach involves harvesting cells from a person, growing them into an assay of organoids, and using testing of treatment responses on these to guide treatment decisions for the same patient. The patient-specific approach is thought of as the most individualised approach, enabling a form of experimentation on a biological "stand-in" of the individual that has never been possible before.

3.3 Main Expectation B: Practical Feasibility

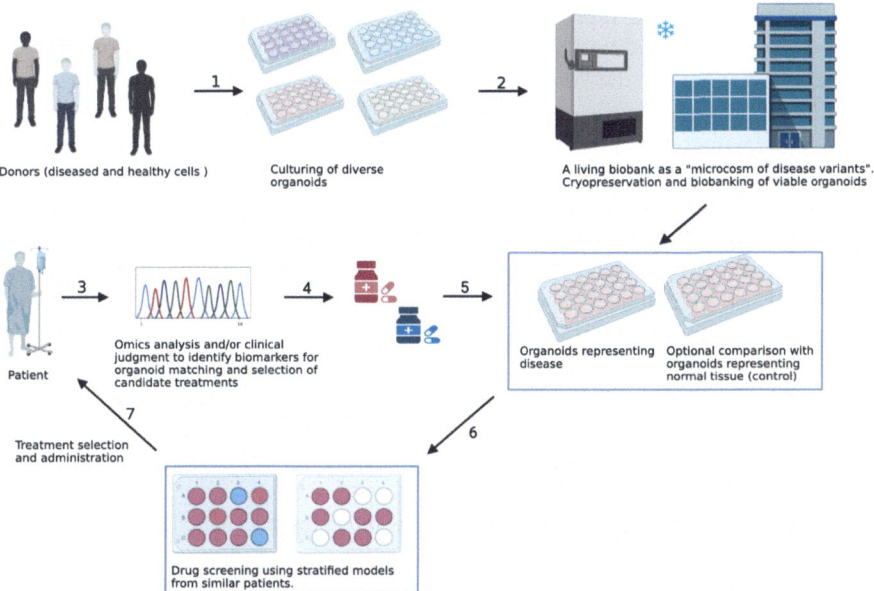

Fig. 3.4 The stratified approach as envisioned, where treatment decisions can be guided by testing on organoids from similar patients. Organoids from diseased and healthy donors can be cultured (1) and cryopreserved at a "living biobank" for future use (2). These can then be used to guide treatment decisions for patients, based on an omics analysis and/or clinical judgement to identify relevant matches (3) and treatments (4) for organoid-based testing (5). The result of testing on organoids from similar patients (6) can then be used to guide treatment selection (7). The figure was created by the authors using BioRender

We can now predetermine a patient's response to any given anticancer therapy by exposing tumor organoids established from the patient's own tumor. This cutting-edge biomedical platform translates to patients being treated with the correct drug at the correct dose from the outset, a truly personalized and precise medical approach. A high throughput drug screen on organoids also allows drugs to be tested in limitless combinations. More recently, the tumor cells that are resistant to the therapy given to a patient were selected in culture using the patient's organoids. The resistant tumor organoids were then screened empirically to identify drugs that will kill the resistant cells. This information allows diagnosis in real-time to either prevent tumor recurrence or effectively treat the recurring tumor. (Vincan et al. 2018)

The use of normal cells of the same person to act as "healthy controls" in the drug screening also opens potential uses of organoids in toxicology testing or testing for patient-specific toxic effects (side-effects) or allergies at the same time. Moreover, if sufficient organoids can be developed, researchers can also test for effects of the combination of multiple treatments.

The most prominent metaphor used to describe these models is that of a patient "avatar", an evocative word well-known from the Avatar movies and computer games.[2] An avatar is a representation of the individual, something that can be "played" with and tested without damage to the real person. It has previously been used as a metaphor for the "digital patient", as virtual computational avatar (Diaz et al. 2013), and also in the context of patient-derived xenograft "mouse avatars" (where mice with transplanted tumours from the patient are considered that patient´s avatar) (Cantrell & Kuo 2015; Francies & Garnett 2015). "In the near future we can envision the ability for oncologists to test patient-derived organoids against a vast range of potential drug options to identify which treatment will be most effective for the patient before treating that patient. The ability to perform such functionalised tests on hundreds of 'patient avatars' will give each patient access to the right treatment, the first time" (Sanctuary 2019).

The patient-specific organoids are also called *"living surrogates"*, meaning a substitute or stand-in (Verduin et al. 2021).

Another tell-tale analogy appearing in a publication by Huch and colleagues (2017) is the antibiogram, comparing PDOs to a long-standing tool in microbiology and infectious disease medicine.

This analogy also emerged in several of our interviews:

> This is very comparable, by the way, to what we do with bacterial infections, where, for the past 60 years, when you have a bacterial infection to your lung, you grow the bacteria in the lab, expose them to all available antibiotics, pick the one that kills the bacterium and this is the one that you give to the patient. (senior researcher, industry)

> This is exactly the same thing that you'd expect if you had a culture taken off your throat – you get a sore throat and they're asking what antibiotic you would use or other sites where we take regularly take cultures. I think that it should be as simple as that. (senior researcher, academic)

> Here we call our test the "chemogram", as compared to the antibiogram, because when you go to a diagnostic lab because, say, you have a pathogenic infection of some sort and it's a bacterial infection, they are not going to try to find which bacteria is the pathogen, they are going just to try to identify which antibiotic works for you. (senior researcher, academic)

This analogy underlines the advantage of testing drugs/treatments on the patient's own cells instead of applying standardised treatments (or instead of testing it on samples from a biobank, see below). Using this analogy communicates that, in some domains, 'personalised' procedures that are easy and highly accessible are on the horizon.

PDO drug testing will become a standard tool when it becomes as simple as an antibiogram, and it "should" work that way, although interviewees referring to this analogy acknowledge that we are not there yet ("it *should* be as simple as that", as obviously it is not). We will get back to this issue in our vision evaluation below.

---

[2] A more recent expression is "three-dimensional patient tumour avatar" *(3D-PTA)*, but this is essentially the same as an "avatar" in the context of cancer medicine (see Bose et al. 2022, Betge and Jackstadt 2022).

3.3.2.2 The Stratified Approaches and "The Living Biobank"

The stratified approaches employ organoids taken from other patients with similar (or sometimes identical) characteristics. The patient is then treated according to results in those organoids. The stratified approach has also been in the vision from the start (see e.g. Francies and Garnett 2015). It is envisioned as superior to stratification based on genomics only.

The stratified approaches rest heavily on the concept of "the living biobank", a concept appearing in the vision from 2015 (e.g., van de Wetering et al. 2015). Living biobanks are evocatively described as "encyclopedias" (Vela and Chen 2015), "repositories", (Vela and Chen 2015; Bose et al. 2021), "libraries" (Mou et al. 2015) of either fully "living" or frozen organoids, preferably numerous, that reflect the variation of a disease between patients (inter-patient heterogeneity). Living biobanks are assumed to form a "microcosm of the disease landscape" (Bose et al. 2021, p. 1014) that provide "ready-made therapeutic response patterns" (Lo et al. 2020, p. 770).

Bose and colleagues summarise the vision as follows:

> Because large-scale PDO libraries can be expanded to include large patient populations, the clinical heterogeneity of various human diseases can be captured in a robust way. As such, PDO biobanks can serve as representative microcosms of the disease landscape as a whole. PDOs of rare disease subtypes can thus be selected and used for drug-sensitivity screening. (Bose et al. 2021, p. 1014)

3.3.3 Prerequisites and Conditions for Practical Feasibility

Organoids are also portrayed as possessing other practical qualities that increase their potential in becoming useful clinical tools. These include the potential of…

- high (reliable) rate of establishment success
- accessing and developing sufficient organoid material, and relatedly…
- …obtaining a big enough number of organoids
- organoids that are produced quickly enough to be used in the clinic
- the ease with which they may be manipulated and their readability when testing treatments
- relatively inexpensive models

These features, if realised, may make organoids practically more feasible than more physiologically complex in vivo models, such as PDX models, i.e., mouse models inserted with tumour cells from specific patients. According to Bose et al. (2021), PDXs have several advantages; "they retain intra-tumoral clonal architecture after repeated passaging, faithfully recapitulate patient drug response, and capture genetic diversity of tumour types across patients" (p. 1018). However, organoids still have a place. The link between intra-tumour mutational heterogeneity and treatment resistance has been demonstrated via single cell sequencing, but this procedure destroys the cancer cells and is typically insufficient to predict treatment response (Sasaki and Clevers 2018; de Witte et al. 2020). The use of tumour organoids, by contrast, allow researchers to expand on different tumour cells and subpopulations to allow for multiple analyses, e.g., comparison of the mutational evolution with histological development and treatment response over time (Mo et al. 2022; Schmäche et al. 2024). While organoids—due to their small size—cannot recapitulate the full tumour, an array of different subpopulations from different tumour segments is hoped to allow for new insights to what difference intra-tumour diversity makes for treatment response (Sasaki and Clevers 2018). Moreover, compared to PDX models, organoids are less time-consuming, less expensive, have higher establishment rates for some cancers, while avoiding ethically problematic use of animal models.

It should be noted that the highlighted practical utility of organoids is also tied to more effective treatment selection. Steuten (2016) pointed out that when assessing the relevance (utility) of a diagnostic technology, we must focus not only on the accuracy of the diagnostic test or ability to predict, but the impact it has on decisions and the effectiveness of the therapies that are selected based on the predictions. This brings us to the question of documentation and amenability for scientific investigation.

3.4 Main Expectation C: Amenability and Documentation

The third main expectation we find beneath the overarching promise, is that the predictive tumouroid models, if useful, can also be documented as such. In our material, documenting that using the models are useful is envisioned as a process where epistemological uncertainty (uncertainty about whether they work) is gradually decreased. Broadly, studies will first document that the models can be valid and predictive; that findings in the models correspond to what one sees in the patients. The second step consists of studies that more definitely document not only such validity, but clinical utility. Here studies where the models actually guide treatment in individuals will be conducted. We will first describe the former kind of knowledge production, then the latter.

3.4.1 Observational Trials (Parallel Trials and Co-clinical Trials)

Observational trials represent an important early step in the clinical documentation process. In these trials, the tumour organoid(s) and the patients are given the same treatments, and one observes if the clinical responses in the patients and the readouts in the organoids correlate. Here, the organoids are not actually used to guide treatment as is the aim for real clinical practice; these studies are observational, not

3.4 Main Expectation C: Amenability and Documentation   45

interventional. These trials are sometimes called "parallel trials."[3] As one of our interviewees states,

> So, the earlier trials with organoids are what we call 'parallel trials.' The patient will go off and take whatever drug is being prescribed by the oncologist, and you put the same drug to the same patient's organoid. You run them in parallel and see if the organoid can recapitulate the response. (senior researcher, academic)

Such studies are also labelled "co-clinical trials," (Nardella et al. 2011) in which cancer organoids are sought provided the same treatment as the donating patient (Francies and Garnett 2015, see also Jeon and Cheong 2019; Bose et al. 2022 and Vlachogiannis et al. 2018 as a prime example).

Tumouroids are envisioned to aid in the evidence production for stratified approaches as described above—or perhaps as evidence to bolster the general idea that the models are predictive when facing unique patients for whom no evidence-based treatments are available. Others see these trials more as proof-of-concept, showing that the models can be predictive, but not that they are clinically useful.

3.4.2 Interventional Trials—"Prospective Validation Trials" and Randomised Controlled Trials

The other main form of clinical trial are prospective clinical trials where experiments on organoids informs clinical interventions (see e.g., Kumari 2020). Such trials may provide generalisable knowledge and are sometimes portrayed as a must for the field.

> As with genome guided therapy, 3D-PTA-guided therapies must undergo rigorous prospective clinical trials to demonstrate clinical benefits for clinical adoption, regulatory approval, and eventually payer reimbursements. (Bose et al. 2022, p. 1451).

One option here is RCTs, often considered the gold standard in evidence-based medicine.

> In the standard approaches we design clinical trials in which we can clearly demonstrate. You could have a trial where you design one arm where you use organoids, the results of organoid testing, to determine treatment, and another using a standard approach, for example the physician preference. (senior researcher, academic)

However, other designs than RCTs are also highlighted. "While the ideal paradigm of the randomised controlled trial (RCT) has long dictated the format of clinical studies, novel study designs may allow PDOs to be evaluated with the same degree of clinical rigor without trial constraints", state Bose et al. (2021, p. 1022). Prospective validation trials, i.e., cohort studies, where clinicians use predictions based on results from the PDO of each patient when choosing treatment, is another option (Bose et al. 2022).

---

[3] Not to be confused with "parallel studies" where different treatments are given to two groups of people.

Adaptive trials are also mentioned as an alternative: "The design of adaptive clinical trials, with a treatment arm allocation according to tumour phenotype and organoid pharmacotype, may be more promising and attractive for both patients and clinicians than traditional RCTs with current treatment standards" (Aberle et al. 2018). Here, information about what happens in patients after the treatment is given is used to guide treatment during the trial, while other studies often fix what treatments patients will receive at the beginning (Schork, 2015).

Both prospective RCTs and "prospective validation trials" are envisioned to…

> …buoy the efforts of early adopters while providing the basis for further clinical utility trials. These studies, with larger cohort sizes, randomized arms, and clinically meaningful endpoints of progression free survival or overall survival, will be the foundation for adoption by the broader community. (Bose et al. 2021, p. 1451)

3.4.3 Single-Case Design Trials and Case Series

The strategies for knowing what works in patients described above are group-based; they study populations large enough to provide statistical support for treating future, similar cases. But in the patient-specific approach described above, an end goal is to treat the patient according to their specific organoid. So, what should be done if statistically based clinical evidence for the specific tumour-drug combination before treating the patient is not available? In these cases, one envisions a combination of using organoids that have been generally validated as predictive using observational trials, and what may be called single-case design studies (SCDs). Such trials can contain various methodological elements and come in a continuum from very rigorous scientific designs to very unreliable designs (Kravitz and Duan 2014; Nikles et al. 2021; Kane et al. 2021). At the most rigorous end of the spectrum, we find n-of-1 randomised clinical trials. Narrowly defined, this refers to a kind of trial with a design where patients alternate between getting the treatment and a control treatment (or placebo) several times to gain statistical material to judge if there is a causal connection between treatments observed in the patient and the treatment (multi crossover design). It also involves randomisation of treatment and control treatment phases and blinding (doctor and patient do not know in which phases the treatment is given). The problem with such trials is that they are costly and often not practically or ethically feasible (e.g., in oncology it is tenuous to withdraw a treatment from a patient when it seems to work).

Less rigorous single-case alternatives are called single case open trials (lacking blinding) and pre-post trials (where there is no multi crossover, but one simply observes the patient before and after treatment (Nikles et al. 2021; Kane et al. 2021).

At the most unreliable end of the spectrum, we find mere case reports (Vogt, 2025 in press). Notably, in single-case strategies, information is generated when the treatment is given to the individual (unlike first generating evidence in population-based studies, and then providing the treatment based on that evidence). A problem with single-case design studies is that they have little or no generalisability; they

3.4 Main Expectation C: Amenability and Documentation

say little predictive about what can be expected in other patients. At the same time, results from SCDs can be aggregated to generate population-based, generalisable knowledge.

Related to SCDs, the organoid field comes with a promise of new forms of evidence production: "Eventually, avatars could create an evidence-based rationale for new clinical trials as this concept has the potential to revolutionize health care process" (Durinikova et al. 2021). This raises the question of how such trials will be organised. Visioneers of the organoid field acknowledge that documentation will not be straightforward. As Bose et al. state, "the path for validation of 3D-PTAs for clinical decision making is yet to be established" (Bose et al. 2022, p. 1451).

Organoids as Models and Diagnostic Tools

To what extent does the clinical utility of organoids hinge on the ability of the in vitro model to be a faithful representation of the patient? Replicability challenges in translational research have often been attributed to the limited similarity between preclinical models and human patients. However, there is no straightforward relationship between representativeness and predictability of models, as also idealised and simple models can be practically useful.

For instance, Knuuttila (2011) argues that looking at models not only as representations of reality, but also as artefacts, is as much as relevant and fruitful if we want to understand their role in scientific practice. As artefacts, that is, engineered entities, models are easier to build, to control, to manipulate, etc., than the more complex model target. All these factors, as well as ethical concerns, enter into consideration when it comes to judge what is a "good" model in biomedical research (Dietrich et al. 2020). There is no "perfect model" of the phenomenon of interest independently of the context of research and of what the model will be used for, and this holds also for organoids (Gaillard and Botbol-Baum 2022).

In our discussion, we also argue that representativeness is not the only characteristic that matters for the tumour modelling enterprise to be successful. This can be rephrased by saying that tumouroids, possibly like all models of scientific research, are instruments or tools. In a clinical context, predictive validity may be more valuable that representativeness. In that sense, PDOs could be closer to a diagnostic tool than a representative model of the tumour. An evaluation of the clinical potential of organoids should therefore not only focus on similarity relations between in vitro model and patient tumour, but also on available evidence to assess their predictive capacities. Hence, our vision evaluation examines both the issue of how organoids relate materially to their in vivo counterpart *and* how their predictive potential is evaluated in clinical trials.

References

Aberle M et al (2018) Patient-derived organoid models help define personalized management of gastrointestinal cancer. Br J Surg 105:e48–e60. https://doi.org/10.1002/bjs.10726

Baker L et al (2016) Modeling pancreatic cancer with organoids. Trends Cancer 2(4):176–190. https://doi.org/10.1016/j.trecan.2016.03.004

Beckmann JS, Lew D (2016) Reconciling evidence-based medicine and precision medicine in the era of big data: challenges and opportunities. Genome Med 8(1):134. https://doi.org/10.1186/s13073-016-0388-7

Betge J, Jackstadt R (2022) From organoids to bedside: Advances in modeling, decoding and targeting of colorectal cancer. International Journal of Cancer 1–10. https://doi.org/10.1002/ijc.34297

Bose S, Clevers H, Shen X (2021) Promises and challenges of organoid-guided precision medicine. Medicine 2(9):1011–1026. https://doi.org/10.1016/j.medj.2021.08.005

Bose S, et al (2022) A path to translation: how 3D patient tumor avatars enable next generation precision oncology. Cancer Cell 40, https://doi.org/10.1016/j.ccell.2022.09.017

Campbell JJ, Davidenko N, Caffarel MM, Cameron RE, Watson CJ (2011) A Multifunctional 3D Co-Culture System for Studies of Mammary Tissue Morphogenesis and Stem Cell Biology. PLoS One 6(9):e25661. https://doi.org/10.1371/journal.pone.0025661

Cantrell M, Kuo C (2015) Organoid modeling for cancer precision medicine. Genome Med 7:32. https://doi.org/10.1186/s13073-015-0158-y

Chakravarti A (2011) Genomics is not enough. Science 334(6052):15. https://doi.org/10.1126/science.1214458

Clara-Trujillo S et al (2020) In vitro modeling of non-solid tumors: how far can tissue engineering go? Int J Mol Sci 21:5747. https://doi.org/10.3390/ijms21165747

Clevers H (2013) A gutsy approach to stem cells and signalling: an interview with Hans Clevers. Dis Model Mech 6(5):1053–1056. https://doi.org/10.1242/dmm.013367

Clevers H (2015) Banking on organoids. Interview by Eric Bender. Nature 521(S15)

Clevers H (2016) Modeling development and disease with organoids. Cell 165(7):1586–1597

de Witte CJ, Valle-Inclan JE, Hami N, Lõhmussaar K, Kopper O, Vreuls CPH, Jonges GN, van Diest P, Nguyen L, Clevers H, Kloosterman WP, Stelloo E (2020) Patient-derived ovarian cancer organoids mimic clinical response and exhibit heterogeneous inter-and intrapatient drug responses. *Cell Reports, 31*(11)

Diaz V et al (2013) Roadmap for the digital patient. Discipulus: European commision

Dietrich M et al (2020) How to choose your research organism. Stud Hist Philos Biol Biomed Sci 80:101227. https://doi.org/10.1016/j.shpsc.2019.101227

Durinikova E, Buzo K, Arena S (2021) Preclinical models as patients' avatars for precision medicine in colorectal cancer: past and future challenges. Journal of Experimental & Clinical Cancer Research 40:185. https://doi.org/10.1186/s13046-021-01981-z

Foo MA et al (2022) Clinical translation of patient-derived tumour organoids- bottlenecks and strategies. Biomarker Research 10:10. https://doi.org/10.1186/s40364-022-00356-6

Francies H, Garnett M (2015) What role could organoids play in the personalization of cancer treatment? Pharmacogenomics 16(14):1523–1526

Gaillard M, Botbol-Baum M (2022) Pursuit of perfection? On brain organoids as models. AJOB Neurosci 13(2):79–80. https://doi.org/10.1080/21507740.2022.2048735

Gao D et al (2014) Organoid cultures derived from patients with advanced prostate cancer. Cell 159(1):176–187. https://doi.org/10.1016/j.cell.2014.08.016

Germain PL (2014) From replica to instruments: animal models in biomedical research. Hist Philos Life Sci 36(1):114–128. https://doi.org/10.1007/s40656-014-0007-0

Gilazieva Z et al (2020) Promising applications of tumor spheroids and organoids for *personalized medicine*. Cancers 12:2727. https://doi.org/10.3390/cancers12102727

Green S, Vogt H (2016) Personalizing medicine: disease prevention in silico and in socio. Humana. Mente 30:105–145

Günther C, Winner B, Neurath MF, Stappenbeck TS (2022) Organoids in gastrointestinal diseases: from experimental models to clinical translation. Gut 71(9): 1892–1908. https://doi.org/10.1136/gutjnl-2021-326560

Hepburn AC, Sims CHC, Buskin A, Heer R (2020) Engineering prostate cancer from induced pluripotent stem cells-new opportunities to develop preclinical tools in prostate and prostate cancer studies. Int J Mol Sci 21(3):905. https://doi.org/10.3390/ijms21030905

Hofmann B (2003) Medicine as techne—a perspective from antiquity. J Med Philos 28(4):403–425. https://doi.org/10.1076/jmep.28.4.403.15967

Hood L, Flores M (2012) A personal view on systems medicine and the emergence of proactive P4 medicine: predictive, preventive, personalized and participatory. New Biotechnol 9(6):613–624. https://doi.org/10.1016/j.nbt.2012.03.004

Hu LF, Yang X et al (2021) Preclinical tumor organoid models in personalized cancer therapy: Not everyone fits the mold. Exp Cell Res 408:112858. https://doi.org/10.1016/j.yexcr.2021.112858

Huch M et al (2017) The hope and the hype of organoid research. Development 144(6):938–941

HYBRIDA D1.3: Gaillard M, Pence P, Botbol-Baum M (2021) The challenging history of organoid research and its implications for ontology and ethics. http://hdl.handle.net/2078.1/272378

HYBRIDA D 2.1: Shoji, J, Davis, R, Mummery, C, and Krauss, S (2022) The research landscape of organoid and organ-on-a-chip models. https://hybrida-project.eu/deliverables/

Iwasaki E et al (2021) Endoscopic ultrasound-guided sampling for personalized pancreatic cancer treatment. Diagnostics 11:469. https://doi.org/10.3390/diagnostics11030469

Jeon J, Cheong JH (2019) Clinical implementation of precision medicine in gastric cancer. J Gastric Cancer 19(3):235–253. https://doi.org/10.5230/jgc.2019.19.e25

Jung P et al (2011) Isolation and in vitro expansion of human colonic stem cells. Nat Med 17:1225–1227

Kane PB, Bittlinger M, Kimmelman J (2021) Individualized therapy trials: navigating patient care, research goals and ethics. Nat Med 27(10):1679–1686. https://doi.org/10.1038/s41591-021-01519-y

Knuuttila T (2011) Modelling and representing: an artefactual approach to model-based representation. Stud Hist Philos Sci Part A 42(2):262–271. https://doi.org/10.1016/j.shpsa.2010.11.034

Kondo J, Inoue M (2019) Application of cancer organoid model for drug screening and personalized therapy. Cells 8:470. https://doi.org/10.3390/cells8050470

Kravitz RL, Duan N (2014) Design and implementation of N-of-1 trials: a user's guide. The agency for healthcare research and quality's (AHRQ) effective health care program

Kumari R (2020) The role of patient-derived tumor organoids in precision medicine. https://blog.crownbio.com/patient-derived-tumor-organoids-in-precision-medicine

Letai A (2017) Functional precision cancer medicine—moving beyond pure genomics. Nat Med 23:1028–1035. https://doi.org/10.1038/nm.4389

Letai A, Bhola P, Welm A (2022) Functional precision oncology: testing tumors with drugs to identify vulnerabilities and novel combinations. Cancer Cell 40 https://doi.org/10.1016/j.ccell.2021.12.004

Lo YH et al (2020) Applications of organoids for cancer biology and precision medicine. Nat. Cancer 1:761–773. https://doi.org/10.1038/s43018-020-0102-y

Mazzocchi A, Soker S, Skardal A (2019) 3D bioprinting for high-throughput screening: drug screening, disease modeling, and precision medicine applications. Appl Phys Rev 6(1). https://doi.org/10.1063/1.5056188

Mo S, Tang P, Luo W, Zhang L, Li Y, Hu X, Ma X, Chen Y, Bao Y, He X, Fu G, Hua G (2022) Patient-derived organoids from colorectal cancer with paired liver metastasis reveal tumor heterogeneity and predict response to chemotherapy. Adv Sci 9(31):2204097

Mou H et al (2015) Personalized medicine for cystic fibrosis: establishing human model systems. Pediatr Pulmonol 50:S14–S23. https://doi.org/10.1002/ppul.23233

Napoli G, Figg W, Chau C (2022) Functional drug screening in the era of precision medicine. Front Med 9:912641. https://doi.org/10.3389/fmed.2022.912641

Nardella C, Lunardi A, Patnaik A, Cantley LC, Pandolfi PP (2011) The APL paradigm and the "co-clinical trial" project. Cancer Discov 1(2):108–116. https://doi.org/10.1158/2159-8290.CD-11-0061

Nikles J et al (2021) Establishment of an international collaborative network for N-of-1 trials and single-case designs. Contemp Clin Trials Commun 23:100826. https://doi.org/10.1016/j.conctc.2021.100826

Ohta Y, Sato T (2014) Intestinal tumor in a dish. Front Med 1:14. https://doi.org/10.3389/fmed.2014.00014

Pfohl U et al (2021) Precision oncology beyond genomics: the future is here—it is just not evenly distributed. Cells 10:928. https://doi.org/10.3390/cells10040928

Riedesser J, Ebert M, Betge J (2022) Precision medicine for metastatic colorectal cancer in clinical practice. Ther Adv Med Oncol 14:1–25. https://doi.org/10.1177/17588359211072703

Sanctuary C (2019) Bringing life to precision medicine. http://www.pmlive.com/pharma_thought_leadership/bringing_life_to_precision_medicine_1280785. Accessed 21.03.2024

Sasaki N, Clevers H (2018) Studying cellular heterogeneity and drug sensitivity in colorectal cancer using organoid technology. Curr Opin Genet Dev 52:117–122

Sato T et al (2009) Single Lgr5 stem cells build crypt–villus structures in vitro without a mesenchymal niche. Nature 459(7244):262–265. https://doi.org/10.1038/nature07935

Schmäche T, Fohgrub J, Klimova A, Laaber K, Drukewitz S, Merboth F, Hennig A, Seidlitz T, Herbst F, Baenke F, Ada AM, Stange DE (2024) Stratifying esophago-gastric cancer treatment using a patient-derived organoid-based threshold. Mol Cancer 23(1):10

Schork N (2015) Personalized medicine: time for one-person trials. Nature 520:609–611. https://doi.org/10.1038/520609a

Simian M, Bissell M (2017) Organoids: a historical perspective of thinking in three dimensions. J Cell Biol 216(1):31–40. https://doi.org/10.1083/jcb.201610056

Singh T et al (2021) Exploring the potential of drug response assays for precision medicine in ovarian cancer. Int J Mol Sci 22:305. https://doi.org/10.3390/ijms22010305

Steuten LM (2016) Early stage health technology assessment for precision biomarkers in oral health and systems medicine. OMICS 20(1):30–35

Sun Y, Ding Q (2017) Genome engineering of stem cell organoids for disease modeling. Protein Cell 8(5):315–327. https://doi.org/10.1007/s13238-016-0368-0

Tonelli MR, Shirts BH (2017) Knowledge for precision medicine: mechanistic reasoning and methodological pluralism. JAMA 318(17):1649–1650

Tutton R (2012) Personalizing medicine: futures present and past. Soc Sci Med 75(10):1721–1728. https://doi.org/10.1016/j.socscimed.2012.07.031

Van de Wetering M, Francies HE, Francis JM, Bounova G, Iorio F, Pronk A, van Houdt W, van Gorp J, Taylor-Weiner A, Kester L, McLaren-Douglas A (2015) Prospective derivation of a Living Organoid Biobank of colorectal cancer patients. Cell 161(4): 933–945. https://doi.org/10.1016/j.cell.2015.03.053

Vela I, Chen Y (2015) Prostate cancer organoids: a potential new tool for testing drug sensitivity. Expert Rev Anticancer Ther 15(3):261–263. https://doi.org/10.1586/14737140.2015.1003046

Veninga V, Voest E (2021) Tumor organoids: opportunities and challenges to guide precision medicine. Cancer Cell 39. https://doi.org/10.1016/j.ccell.2021.07.020

Verduin M et al (2021) Patient-derived cancer organoids as predictors of treatment response. Front Oncol 11:641980. https://doi.org/10.3389/fonc.2021.641980

Vincan E et al (2018) The central role of Wnt signaling and organoid technology in personalizing anticancer therapy. In: Larraín J, Olivares G (eds), Progress in molecular biology and translational science, vol 153. Academic Press, pp 299–319 https://doi.org/10.1016/bs.pmbts.2017.11.009

Vlachogiannis G et al (2018) Patient-derived organoids model treatment response of metastatic gastrointestinal cancers. Science 359(6378):920–926. https://doi.org/10.1126/science.aao2774

References

Vogt H, Hofmann B, Getz L (2016) The new holism: P4 systems medicine and the medicalization of health and life itself. Med Health Care Philos 19:307–323. https://doi.org/10.1007/s11019-016-9683-826821201

Vogt H (2025) Personalized medicine beyond stratification. In: Schramme T, Walker MJ (eds), Handbook of the philosophy of medicine, 2nd edn. Springer. In press

Wang Y, Li Y et al (2022) Advances of patient-derived organoids in personalized radiotherapy. Front Oncol 12:888416. https://doi.org/10.3389/fonc.2022.888416

Xu R, Zhou X, Wang S, Trinkle C (2021) Tumor organoid models in precision medicine and investigating cancer-stromal interactions. Pharmacology & Therapeutics 218: 107668.

Zanoni M, Cortesi M, Zamagni A, Arienti C, Pignatta S, Tesei A (2020) Modeling neoplastic disease with spheroids and organoids. Journal of Hematology & Oncology 13:97. https://doi.org/10.1186/s13045-020-00931-0

Open Access This chapter is licensed under the terms of the Creative Commons Attribution 4.0 International License (http://creativecommons.org/licenses/by/4.0/), which permits use, sharing, adaptation, distribution and reproduction in any medium or format, as long as you give appropriate credit to the original author(s) and the source, provide a link to the Creative Commons license and indicate if changes were made.

The images or other third party material in this chapter are included in the chapter's Creative Commons license, unless indicated otherwise in a credit line to the material. If material is not included in the chapter's Creative Commons license and your intended use is not permitted by statutory regulation or exceeds the permitted use, you will need to obtain permission directly from the copyright holder.

Chapter 4
Vision Evaluation of Patient-Derived Organoids for Personalised Medicine

We now come to the actual critique of the vision. Having laid out the vision for evaluating in the previous chapter we can delve into the "main dish" in our argument. In our vision analysis we described the overarching promise of the vision as being able to *improve predictions of what the patient with cancer disease needs and provide significant clinical utility through personalization of treatment*. Our evaluation of this promise here is multifaceted and used all our materials. We address the main expectations of the vision, and the evaluation of each of these could in terms of length have been printed as chapters of their own. We have, however, chosen to keep them in one chapter, as they form parts of a whole evaluation.

We will argue that in terms of hype and plausibility, the bolder versions of the vision seem implausible. However, the more moderate versions seem more realistic. We point out limitations and challenges at many levels of this vision, focusing on both model uncertainty (differences in what models need to capture of reality and what they likely can capture), uncertainty concerning the published literature and planned research, and uncertainty about how the vision can be documented.

4.1 Uncertainty Regarding Main Expectation A: Representation (Model Uncertainty)

First, in this chapter, we critique the idea that organoid models can more accurately capture the true complexity of a patient's specific tumour as well as variation between individuals (inter-tumour heterogeneity) than other models, and that they therefore will give better predictions for the individual.

This discussion revolves around *model uncertainty*, i.e., the kind of epistemological uncertainty that is linked to the discrepancy between the model and the real-world phenomenon it emulates (Djulbegovic et al. 2011). "*All models are wrong, but some are useful*", is an insight often attributed to British statistician George Box.[1] The question here then, is *how wrong* the organoid models are, and how this wrongness impacts the fulfilment of the vision. How "physiologically relevant" can these "disease in a dish" models be? In the following, we rely first on limitations and challenges that practitioners in the organoid field state themselves, and then we raise questions that came up during our own critical analysis.

In structuring the many elements of this critique, we follow the way these models are designed and developed over time. We follow here central theorists within the field who very clearly state that the issues that limit the translatability of organoids (i.e. their implementation and usefulness in the clinic) "are inherent to their very design principle" (Hofer and Lutolf 2021), and that the challenges "can be divided into different phases of the organoid workflow process" (Bengtsson et al. 2021). We will start with the beginning of an avatar's "life", the phase that the field calls "establishment".

4.1.1 Issues with the Harvesting and Processing of Cells

In developing a cancer organoid, there is first a need for cells. These cells must be representative of the tumour—and only the tumour. This presents a huge challenge in the form of intra-tumour heterogeneity: Tumours contain cells of different types, with different mutations, and different responsiveness to different medications. Getting a representative "catch" of this intra-tumour heterogeneity or "neoplastic cellularity" from the beginning is important (de Witte et al. 2020; Bengtsson et al. 2021).[2] However, as one author in the field states, there are challenges:

"One major bottleneck is that tumour organoids are often derived from biopsies representing only a small part of the entire tumour. In this way, the complexity of the original malignant lesion will always be underrated, and intra-tumoral heterogeneity could hinder clinical translation" (Wang et al. 2022). The quote points to the risk of harvesting tumour cells that are not fully representative of how the tumour, as a whole, develops and reacts to treatments. Different cells or subpopulations of tumours may need different types of treatments, and ongoing organoid research therefore investigates differences in organoid predictions depending on whether the cells are harvested from primary or metastatic tumours, or from multiple biopsies at the same time (de Witte et al. 2020; Mo et al. 2022; Schmäche et al. 2024).

Another problem is that tumour organoids can be considered as "living" models of a moving target, as cancer cells develop over time both in vitro and in vivo.

---

[1] https://en.wikipedia.org/wiki/all_models_are_wrong (accessed 21.03.2024).

[2] To what extent do the cells need to represent the whole tissue? There is uncertainty on what it takes to achieve sufficiently representative models for the purpose of prediction. Models have to be representative only in so far as representativeness is key for a better prediction. Depending on the cancer type and nature of the tumour, prediction from a single cell might be an option. This issue of intra-tumour diversity and the ability of organoids to account for this is currently explored by organoid research (Mo et al. 2022; Schmäche et al. 2024).

4.1 Uncertainty Regarding Main Expectation A: Representation (Model ...   55

A pressing question is to what extent the artificial growth conditions of organoids affect the representativeness and predictability of organoid models. As described elsewhere, tumour organoids are simplified models compared to in vivo tumours, and they do not only contain patient-specific cells. "Currently, there is no good method to obtain pure tumour organoids", writes Chen et al. (2022). Normal epithelial cells or cells from other animals from culture media may be part of the organoid and influence its growth. Obtaining such purity is acknowledged as a crucial challenge (Pang et al. 2021). Additionally, there are problems with how harvested tissues are handled, minced, separated and perhaps frozen and stored in biobanks, as induced differences and contamination of cell cultures can occur (LeSavage et al. 2022). All these initial conditions of organoid growth affect how well organoids represent the relevant aspects of the disease in the patient.

4.1.2 Issues in Accounting for the Environment

Here we touch the vision's key promise of representing not only the cells and genome (DNA) of the tumour, but their environment, the context. We are here talking specifically about the ability of the models to represent the initial growth environment of the tumour organoids. Just like the early environment in the womb and after birth is crucial for the development of the organism, the initial artificial tumour microenvironment (TME), is crucial to organoid development. It guides and constrains development and determines the resulting model.

The environment in which stem cells grow into organoids is an extracellular matrix (ECM) consisting of different proteins in a solution. The most commonly used ECM products are basement membrane extracts (BMEs) with the brand name Matrigel®. In the field of organoids, there is a strong, explicitly stated consciousness about not being able to provide the growing cells with a growth environment that is representative of the actual conditions of the body. This may be called "*The Matrigel® Problem*" (see e.g., Pamarthy and Sabaawy 2021; Zhou et al. 2021). This problem concerns limitations of this growth medium, both in terms of lack of standardisation (see Sect. 5.2) and accuracy in the representation of real-life conditions. Matrigel® consists of proteins from the extracellular matrix and basement membrane of mouse sarcomas (a form of cancer arising in muscle and connective tissue).[3] The problem is, in the clinical context of PM, that "avatars" of humans will here have to start their "lives" in a solution which is taken from another animal with a specific cancer. This creates both known and probably unknown uncertainties about the accuracy of the models. One of our interviewees raises the concern:

> It's what almost everybody doing organoids uses, but Matrigel material that allows for the 3D culture of the organoids is derived from mouse sarcoma tumors, and so, in my opinion, if you use Matrigel you're already biasing the micro-environment around your cells towards

---

[3] According to LeSavage et al. (2022), the same issue arises from the widespread use of foetal calf serum in growth media.

mouse sarcoma… I think that there needs to be a shift away from Matrigel, because that could, across the field, bias all the results, including all the clinically oriented results, which could be extremely problematic. I think that if the field keeps using that material, and we keep doing this for 25 years, we might have created this technology that is actually less accurate or representative of human biology than we think it is. (senior researcher, academic)

The ECM may also contain unknown or "undetermined extracellular components, which may unexpectedly modify biological cell behavior" (Zhou et al. 2021). This will "make it impossible for organoids to fully simulate the physiological function of organs" (or tumours) (Lin et al. 2022), something that will "hamper in vivo comparisons, and also impair drug penetration with subsequent detrimental effects on the utility of organoids" (Clark et al. 2022).

Importantly, just what is the correct growth medium is itself an unknown. The question of making the right culture medium is thus not just about making something we know the perfect composition of, but fundamental uncertainty about what the aim should be (Wang et al. 2022a, b, c).

Cells in an organoid develop in the absence of—or at least very different—physical interactions with the environment that we find in the body also when it comes to physical tensions, mechanical stresses (such as peristalsis in the gut, or respiration movement in lungs) and microscale flows. This also affects the accuracy of the model. Zhou et al (2021), for example clearly state that "…current cancer organoid cultures do not replicate accurate mechanical control and physical manipulations that occur in vivo".

There is additionally an issue of lack of knowledge about the various factors that can influence organoid development outside the human body. Le Savage et al., writing about cancer, states that, "unfortunately, for many of these cases, the specific mechanisms mediating the successful establishment of organoids from select tissue samples over others is not well understood" (LeSavage et al. 2022). In sum, the initial growth environment in which the harvested stem cells start to grow and that guides their development, is not as accurate a representation of the diversity of conditions for human cancer as one might hope. This will affect how representative the models are in mimicking responses to treatments. This is not only a technical problem, but a fundamental lack of knowledge.

After the very first phase in organoid development, there are further issues that are related to the promise of representing the (micro)environment of the cells in the patient's body (Hofer and Lutolf 2021).

One critical limitation that impacts the accuracy of organoids is the lack of nerves, vascular system (blood vessels) and immune system (Podaza et al. 2022). It fundamentally limits the complexity and the accuracy of the models (Baker et al. 2016; Hofer and Lutolf 2021). Some researchers call organoids "simplistic" and "reductionist" for this reason (Letai et al. 2022, Chumduri and Turco 2021). It represents a significant breach of the promise of accounting for the microenvironment of cells in the patient's body. Organoids can only "partially simulate the disease process" (Ma et al. 2021). The consequence could be limitations for their predictive power and utility (Luo et al. 2022; Letai et al. 2022; Clark et al. 2022).

4.1 Uncertainty Regarding Main Expectation A: Representation (Model ...   57

> Most PDOs lack stromal cells. Therefore, they fail to reconstitute the microenvironment, which includes fibroblasts, endothelial cells, immune cells, and ECM, and lack the signaling that prompts organogenesis, hindering the ability to accurately predict clinical outcomes. (Qu et al. 2021, p. 1347)

As a particularly vexing problem for cancer, lack of immune cells hinders the testing of immune therapy drugs, which are at the forefront of today's oncology (Letai et al. 2022). As Xu et al. (2022) summarise: "the current organoid technology is unable to easily replicate the complexity of the patient-specific immune environment (…) Imprecise modelling of the tumour immune environment prevents organoids from being useful for translational medicine and precision medicine".

Although far less of a focus than the microenvironment, another fundamental issue is that organoids lack a representative *macroenvironment,* the context that is outside the immediate environment of cells, but still influences it greatly. No organ is an island, and the lack of signals from other organs is bound to affect both the accurate development of the model and the way in which the model can represent an organ or tumour at any given time. Hormone levels, nutritional status of cells and epigenetic factors are all hard to mimic (Sun and Ding 2017).

For instance, an environmental factor that has received widespread attention in the context of modelling gastric cancer is a lacking microbiome, the sum of all bacteria that colonise and interact with the body, particularly in the gut. Poletti et al., for example, state that… "maintaining the gut microbiota and host cells in a co-culture system is currently challenging" (Poletti et al. 2020). This is thought to be significant in the field, particularly since gastrointestinal cancer development is affected by microbiota (Kiwaki and Kataoka 2022).

As Boers et al. (2016) noted: "Although the organoid model closely mimics the dysfunction of the original organ, it does not account for an entire body, or for the broader context of the patient" (p. 940). Although it is not mentioned in our material, human beings, and therefore all organs, and tumours, stand in interaction with a very complex outside environment, including a social environment. Mediated by the brain, and how we interpret the world, this social environment also affects the macro- and microenvironments in the body (McEwen and Getz 2012). These are the interactions highlighted in the biopsychosocial theory that underpins much of modern clinical practice (Engel 1977). Organoids cannot mimic such biopsychosocial interactions. Although organoids are useful precisely because they are simplified models of human parts, it does mean that using them as "avatars" for humans in medical decision-making risks not representing important factors in human health.

These factors combined amount to serious problems in fulfilling the central promise of accounting for the environment. Although most of these limitations are recognised by the field and are actively researched (except representing the wider biopsychosocial situatedness of humans), the organoid field can be said to have a central problem here.

4.1.3 Self-organisation: Issues with the Main Design Principle

The shortcomings of organoid technology in accounting for the context is in turn linked to what is described as the main design principle of organoids: That these models to a certain extent can "model themselves" without too much external control on the properties of stem cells and the process of self-organisation. This capacity, it turns out, is both the beauty of organoids, but also one of their major limitations. Many authors in the field highlight the way too little control of the developmental process, especially growth conditions, represents a "ceiling" against which the ability of organoids to represent reality budges (Hofer and Lutolf 2021). Without control, the self-organising process becomes "unlawful", and perhaps more of a chaotic and partly disorganised system, instead of an organised and complex one (Moreno et al. 2011). Such criticisms are particularly coming from the organ-on-a-chip field, which seeks superior models precisely by providing such control:

> ...these models mostly rely on self-organization of cells with limited control over the tissue culture microenvironment, which means that many aspects of tissue function like transport, metabolism, inflammation, vascularization, contraction, sensing and signalling, are challenging to recreate in a controlled and patient-specific manner. (Van den Berg et al. 2019, p. 199)

This has consequences for the accuracy of organoids, how they represent reality in a valid manner, but also for the technical variability of results (see below).

Regulating self-organisation to generate developed organoids is thus seen as a crucial need in improving personalised models (Shariati et al. 2021). However, again, the solution to this problem is not just a technical one. It is a fundamental problem that also hinges on lack of biological understanding: "...how stem cells self-organise remains obscure. It is therefore often difficult to control the cell type, cell organisation, cell–cell and cell–matrix interactions" (Sun and Ding 2017). The beauty of organoids is that they to a certain extent model themselves, but where that capacity trails off and humans have to take over, we lack knowledge. The more we engineer the process, we also may lose the capacity of organoids to self-organise. The balance between self-organisation and engineering organoid development is therefore a tricky one.

In sum, the assumption that organoids may self-organise to become representative "avatars" of individual patients, is partly tenuous.

4.1 Uncertainty Regarding Main Expectation A: Representation (Model ...

4.1.4 Issues with Size, Maturation and Long-Term Culture

4.1.4.1 Problems with Size

Organoids without stromal cells, in particular vasculature (blood vessels), are limited in size. Like all living tissues, organoids need life-critical factors like oxygen, nutrients and other molecules important for metabolism. Without a blood supply, they rely solely on the culture medium for these factors, but they can only reach so far by diffusing ("leaking") through the tissue. As a result, the organoid's central cells become hypoxic and then die when it grows beyond a certain size, resulting in necrosis (see, e.g., Hofer and Lutolf 2021). Organoids start to suffer and thus behave inaccurately at the size of a tenth of a millimeter and inner cell death occurs at a half a millimeter.[4]

In some instances, size and weight matter in the real world, and this becomes a limitation in accurately representing disease. For example,

> Tumour size at diagnosis is frequently used to estimate prognosis. Larger tumours are often correlated with increased metastatic risk. This can be attributed to the fact that given a certain mutation rate, size becomes a key factor in predicting the presence of drug-resistance mutations. (Foo et al. 2022, p. 5)

4.1.4.2 Problems with Genetic Representation Over Time

As shown in our vision analysis, a critical assumption is that the organoid retains the same genome as the tissue or tumour from which it was derived, and therefore serve as good genetic representations. However, this assumption may not hold over time, especially in the case of cancer. A question here is *genetic drift*, that some genes become more common in the organoid over time by random chance as cells divide and mutate:

> Although most tumor PDOs recapitulate the genetic composition of the parental tumor at early passages, the extent of genetic drift or the proportion of genetically stable cells in organoids at later passages has not been fully characterized (21). (Zhou et al 2021)

One of our interviewees also states:

> Having an organoid line that has been passaged 20 times, which is around a year, in a dish, to what extent does it still resemble the original tumor? And it's completely biological, right? It's completely inherent to cancer biology that the genome evolves, but it evolves under a pressure that is not in the body, so it evolves in a way which is most optimally fit to grow in a dish. So, having these very long-term cultures, you are going to develop something that is not representing what was in the patient. (junior researcher, industry)

---

[4] According to Sun et al. (2022), "The absence of vasculature limits the size of the organoid, and organoids larger than 100–200 μm in diameter suffer from the diffusion of nutrients, oxygen, and metabolites to the central region of the organoid. Furthermore, organoid fragments larger than 500 μm diameter show necrosis ". A micrometer (μm) is a millionth of a metre.

4.1.4.3 Loss of Heterogeneity

Related in part to this genetic drift, there may be an accumulation in organoids of cells that are more successful in reproducing than other, which "die out" in an evolutionary sense. This threatens the models' ability to recapitulate the real heterogeneity of the original tumour:

> With extensive passaging of organoids resulting in loss of heterogeneity due to cellular adaptation to culture conditions in vitro, it raises concerns as to whether tumours organoids are able to accurately recapture ongoing disease process in patients for the purposes of real time monitoring of drug response. Whether repeat biopsies of tissue should be taken and whether tissue would be sufficiently available after initiation of treatment would need to be considered as well. (Foo et al. 2022, p. 12)

Both genetic drift and loss of heterogeneity has consequences for the accurate representation of organoids and accurately modelling phenomena that are time dependent.

4.1.5 Towards More Holistic Organoid Models of Higher Complexity

The fact that organoids struggle in their complexity and representation of the microenvironment, has resulted in promises of making the models more complex and "holistic". Organoid visioneers are looking for ways to introduce—or "co-culture"—other components into organoids. This is sometimes presented as a form of "holism".

> Nowhere has the need for holistic tumor microenvironment culture been more acute than for tumor immunology and the accompanying need to study interactions between cancer cells and their veritable ecosystem of co-habitating immune cells. However, conventional organoid models of cancer have typically only represented tumor epithelium, and holistic culture of cancer cells alongside endogenous stromal elements has been elusive. (Lo et al. 2020)

In one form of approach to this, autologous (meaning coming from the same person) immune cells are co-cultured or added to the organoids after they have been generated, and in another, the other cells are sought included in the culturing of the organoid from the beginning, e.g., air–liquid interface (or ALI) models (Lo et al. 2020).

There are also discussions about how the physiological features of organoids could be improved in the future, but it is unclear to what extent this can be done:

> Although it has been reported that tumor organoids are co-cultured with immune cells or fibroblasts to simulate the tumor microenvironment, compared with the real physiological environment, the composition of organoids is still relatively simple. How to effectively introduce blood vessels and nerve cells into the culture system is also a major problem. (Lin et al. 2022)

This vision evaluation gives a picture of the limitations as perceived around 2020, and we refrain from going into further details about approaches to make organoids "less reductionist" and more "holistic" here. However, it is currently an interesting question to what extent complex diseases can be represented in vitro, and what benefits and challenges are associated with more complex models. This perceived need to build more complex, more "holistic" models with more control over different parameters is also a central impetus for the vision of organ-on-a-chip for PM (Van den Berg et al. 2019).

4.1.6 Simplicity Versus Complexity: The Accuracy-Reliability Trade-Off

As seen in the previous sections, a recurring question concerns the relationship between accurate representation of models and their predictive capacities. It is often assumed that more complex models with more accurate representation of a target will give better predictions, but this is not always the case. Sometimes, more general models with less detail are more predictive. It is not necessarily problematic for a predictive model to misrepresent reality as, "in order for the model to meet its predictive purpose, it needs to be similar only in the way that it relates model inputs (particular drugs or combinations in particular doses) to outputs (treatment response)" (Walker et al. 2019, p. 144).

Importantly, it is not only accurate representation that influences the predictive capacities of a model or a medical test. It is also the precision (also called "reliability"). This is the ability of the test to produce the same result every time, which is also related to the repeatability and reproducibility of results produced by the models. There is a well-known trade-off in medical tests and models between complexity and precision. In the context of organoids, the models might produce unreliable results if there are high levels of variation in how they have been produced. More complex organoids have more factors that may produce technical variability. One author, for example, notes how there must be a "trade-off between complexity and physiological relevance with reproducibility, ease of use and cost" (Clara-Trujillo et al. 2020). Models that are complex enough to represent human biology is a boon, but simplicity is also a condition for useful models that is in tension with this goal. Organoids may thus reach a better balance between representation and reduction of complexity, where selected and clinically relevant features are recapitulated in as simple models as possible. But this is still an open empirical question.

4.2 Uncertainty Regarding Main Expectation B: Practical Feasibility

We now move on to critique the second main expectation we described in the vision analysis: The expectation B, that it will be practically feasible to use these "personalised models" in the real-world clinic so that they are useful in practice.

For this expectation be fulfilled, the field must overcome hurdles that range from the technical and scientific to practical issues in the social realm of the clinic. Some of these issues are not easily separated from the fundamental biological issues we discussed under main expectation A. But some are also obviously of more practical nature. We begin this section by continuing to focus on the precision or reliability of the models (and relatedly the reproducibility of results). While we there focused on a fundamental trade-off between complexity of the models and precision/reliability, we here focus on a range of technical factors that contribute to variability (noise) in test results.

We will first consider problems pertaining to the "patient-specific approach" that we described in the vision analysis, and we will then consider the "stratified approach" relying on biobanks as a proposed solution to some of these problems.

4.2.1 Lack of Standardisation and Technical Noise

Modelling in precision medicine should ideally capture and reflect biological, real-world, variability (or heterogeneity) both within and between patients but avoid variability that is due to chance or technical differences between each model. A critical, practical problem for the implementation of organoids in the clinic is a long list of technical factors that is linked to lack of standardisation and control of the design process (growth) of organoid models. These all lead to variability in the properties of the models and thus the results of tests performed on them that is not related to the real-world variability between patients. This variability (or unreliability) is also called *technical noise*. It leads to less predictive capacity (higher total error) and usefulness of the models. Organoids are complex structures that require complex developmental processes, which means that there are numerous variables that can be non-standardised. The problem of technical noise appears on different levels: Between different organoids produced from the same patient, between patients in the same lab/clinic, and between labs/clinics and studies. It arises from differences in methods, the materials that are used, and human skills, (Luo et al. 2022). From the beginning, as Hofer and Lutolf (2021) state, "the initial conditions of organoid growth contribute to organoid variability, including the starting cell population, their positioning and aggregation."

In the harvesting of cells from the patient, the collection of a tissue sample is non-standardised, creating variability in the composition of the cells that are taken out and grown (Bose et al. 2021). Thus, even from the same patient, the organoid(s)

4.2 Uncertainty Regarding Main Expectation B: Practical Feasibility

that are developed may not be alike. There may also be variability between patients because they have different histories with treatment, even though they have the same disease.

Variability also arises in the handling of cells and organoids at every step, in separation of cells, mincing, freezing, and storage causing lack of reliability among results. Critically, the growth medium shows variability between products, and within different batches of the same product ("batch-to-batch variability"), which contain many substances that each and in interaction may influence growth (Schutgens and Clevers 2020; Zhang et al. 2020; Günther et al. 2022). This is another aspect of "The Matrigel Problem" discussed in Sect. 4.1. Growth media have up until now been ill-defined and contain undefined factors that vary, and they may be mixed and used in non-standardised ways. As one of our interviewees states,

> That is what needs to be resolved when it comes down to – not so much the method anymore – but when it comes down to the materials, right? If you look at the organoid build-up materials, if you see using it industrially, you will see big, black boxes in the built of materials. Where you have: 'oh we just used this' or 'we decided to put in this media component what we got from Becton Dickinson' where in the US they get it from Becton Dickinson as well, but one was sourced in China and the other one is sourced in Romania, you see, and then it's a big mess. We are looking at supply chains all over the place and that's the beauty of GMP, right? When you are working in GMP, it is standardized, super-expensive but that is the big hurdle that would have to be overcome. (senior researcher, industry)

Importantly, there is also lack of knowledge about what the growth medium should ideally be like.

> There have been no validated culture mediums for establishment of tumour organoids and researchers made modifications (…). These adjustments to the culture media are based on 'informed guesses' by the researchers which may have negative impacts on tumour organoid growth. (Foo et al. 2022, p. 12)

Moreover, the differences in physical strain, stress, weight and manipulation during organoid development may cause technical variability. For example, Luo et al. (2022), write that "uniaxial strain induces the growth and maturation of human intestinal organoids". All of these factors influence the self-organisation process of organoids. However, more fundamentally, the self-organisation process—the major design principle of organoids—is in itself a stochastic (random) process when it is not tightly controlled: "Organoid systems also suffer from considerable variability in organoid formation efficiency, end-point morphology and function, which is often inherent to the stochastic nature of in vitro self-organization and cell fate choices" (Hofer and Lutolf 2021). In other words, organoids can vary even when the growth context is the same. During the maturation of organoids, the random genetic changes that take place during cell division, can create variability among organoids: "Genetic heterogeneity leads to different gene expression patterns, which leads to different drug sensitivity" (Luo et al. 2022). Importantly, the growth medium may influence mutations in tumours.

> To add on, studies have also shown that medium composition exhibits selective pressure on PDTOs and may influence the genetic composition of cancer organoids via selection against certain tumour subclones. (Foo et al 2022, p. 12)

The consequence of all these sources of technical variability (noise) is severe. It means that the total error of results of tests performed on organoids goes up and they become less predictive. As one of our interviewees sums up:

> I think the challenge of moving this to the clinic is about the huge challenge in standardizing. Every person, every lab you go to, all around the world, they all claim to have organoids, but every single one has a different protocol. Everyone has sources from different suppliers, some people make their own products, some people don't, some people supply them, then you find confirmed suppliers like Corning... Batch-to-batch variability is ridiculous. For the moment, everything is set up nicely to work even in a sub-good laboratory practice. I wouldn't even say it could be used for a confirmed medically approved diagnostic tool yet, in the next, even, five, eight years. Just based on the regulatory hurdles: when I say that, there is a checklist of everything that needs to be confirmed and organoids from a product, from an application, from a reproducibility viewpoint – it doesn't fit. (senior researcher, industry)

In sum, "the limitations are today the fact that the organoid technology is still not standardized and scalable. Can we make technological developments that would help us to propose this kind of technology for cancer patients at large?", asks another of our informants (senior researcher, academic).

The widespread realisation in the field, which is reflected in our material, is thus that virtually every step in the generation of organoid models for PM must urgently be standardised and more tightly controlled (Luo et al. 2022; Mackenzie et al. 2022; Kiwaki and Kataoka 2022): "To make further progress in organoid technology, all materials, protocols, and quality control procedures used to establish organoids must be completely defined and standardized" (Qu et al. 2021). This includes decreasing the human element in their creation (Bose et al. 2021). As one of our interviewees states:

> The current version of the assays happens in basic research labs, so it's expensive, it takes weeks... There are actually several companies that build these machines that do this very fast with microfluidics in tiny volumes. In about 7 to 9 days you can, from a live sample of the tumor, run it through the machine, test again hundreds of combinations of drugs, for about 300 to 400 dollars and using a machine in a standard diagnostics lab that a technician can run. That is what is needed now, if this is going to be accepted and validated. You need to have an automated version, with tiny volume of samples, low amount of cells, yet with great sensitivity. (senior researcher, industry)

However, this is far from straightforward: "3D organoids generation combines differentiation of stem cells, aggregation of multi-types of cells and diverse bioengineering approaches, making it extremely difficult for standardization and hard for groups to compare results between systems" (Sun and Ding 2017). Notably, self-organisation in itself cannot be standardised, although the developmental process can be more tightly controlled, which precisely becomes the ambition of the OoC field (Van den Berg et al. 2019; Hofer and Lutolf 2021).

4.2.2 Establishment Success Rate: Problems with Generating Organoids

For organoid "avatars" to be a reliable strategy for helping patients in the clinic, one must be able to establish them in a high proportion of patients, that is, have a high success rate. At this point in time, however, the establishment success rate is low in many cancers, and it varies greatly (Letai et al. 2022). As one of our interviewees underscores, this is a significant problem for practical feasibility today:

> If you look at cancer, if you took 100 patients and depending on the type of disease they have, you may be able to get anywhere from 60 to 70% of the patients deriving organoids – you might be able to get organoids from 70% of patients down to as low as 5%. If you have an approach, it obviously should be amenable to all patients, and that is not where we are today. (senior researcher, academic)

This is less of a problem where one has more time and more tissue material to grow the organoids from, but it may be where tissue biopsy is difficult or dangerous, such as brain cancer.

There are many reasons for low establishment rates, connected to the methodological problems we have visited above. One problem is simply the skill of lab workers, as these are complicated processes, and the lack of standardised, optimal procedures (Foo et al. 2022).

More fundamentally, there is also the issue of getting enough material, enough cells, from the patient when establishing the organoid(s) (Clark et al. 2022; Kiwaki and Kataoka 2022; Letai et al. 2022; Foo et al. 2022). "If only low cell numbers are collected, then the establishment of the PDO models becomes very difficult" (Wang et al. 2022a, b, c, p. 6). The size of the starting material, and the number of cells with stem cell properties and ability to divide is important. Therefore, resections of cancer tissue are in one way preferable to smaller needle biopsies and circulating tumour cells for example. "In this context," Li and Selaru (2022) write, "it should be noted that current clinical grade diagnostic protocols are based on needle biopsy and not on surgically resected specimens".

According to our material, this seems especially to be a problem with metastatic cancer, where needle biopsies of metastatic lesions often yield limited numbers of cells. According to Foo et al. (2022), "This would likely translate into the comparatively lower establishment rates of tumour organoids from needle biopsies of metastatic sites as opposed to primary tumour" (p. 11). This is a significant practical problem as many of the patients for whom the development of an organoid avatar is clinically indicated, at least at this point, are cancer patients with metastatic disease who have exhausted other treatment options.

At the same time, it is noted in our material how only cancers with cells that divide relatively quickly (high cancer cell proliferation rate) are readily grown. This means that cancers that are aggressive can be most easily cultured (i.e., poorly differentiated cancers). This might mean "excluding the possibility of deriving tumoroids from patients at early stages of the disease" (Caiazza et al. 2021, p. 8).

Additionally, neoadjuvant treatment (drugs that are used before surgery) can make the tumour smaller, give fewer cells in biopsies, and cause cancer cells to be less likely to proliferate and thus affect success rates (Yan et al. 2022).

Overgrowth by other cells in the organoid is also another main factor behind low establishment success rates (Rossi et al. 2022; LeSavage et al. 2022). Normal epithelial cells grow faster than cancer cells in culture, as cancer cells are prone to error in cell division (Lin et al. 2022). This is a problem that especially hampers the success rates of prostate and lung cancer organoids, both of which are among the most common cancers (Bose et al. 2021).

In sum, the strategy of making "avatars" for patients does not work for everybody. That may not be a requirement, and the strategy may first be directed at specific cancers and diseases with high establishment rates. However, it is presently unknown, in how many it will work. The consequence of low establishment success rates, is, as Foo et al. (2022) state, that "it will not be feasible to recommend this technology for cancer patients who require reliable answers in a timely manner. Until there is an increase in success rates of establishment, it would be a challenge for tumour organoids to be used in clinical practice" (p. 12).

4.2.3 Generating Enough Organoids for Testing

Related to success rate is the problem of generating enough organoids for each patient to conduct validated drug screenings. In some cases, more than one drug needs to be tested, and it is itself a question of how many organoids one needs to test one drug or a combination of drugs. As one of our interviewees states:

> One of the other technical issues – it's a little bit logistical too – is just throughput: how many organoids do you need to produce for a given patient to run the studies that you need? In some types of cancers, the tumor that you take out surgically is big, so you can make hundreds and hundreds of organoids; in other cases, it might not be a surgery, but it might be a needle aspirate where you get only a tiny amount of tissue and you are very limited in what you can do with that. (senior researcher, academic)

It should here be stressed that several of the highlighted potentials of organoids rest on the possibility of having enough—and sufficiently representative—tumour material to work with. For example, the ability of organoids to account for intratumour heterogeneity (Sect. 3.3) is dependent on availability of sufficient material from different tumour segments to compare sub-clonal cell populations. While this is a possibility for some cancers, especially if primary tumour tissue is available, this is much more challenging for surgically inaccessible tumour and metastatic cancer.

4.2.4 "At the Right Time"? Challenges of Speed in Growth

For many cancer patients, time is of the essence. They may become too ill to receive (experimental) treatments or even die before results from organoid drug screenings are available (Shiihara and Furukawa 2022; Foo et al. 2022). As Verduin et al. (2021) state,

> ...the derivation time of most cancer organoids is currently still weeks to months. If cancer organoids were to be used as co-clinical avatars this derivation time needs to be shortened to be of actual clinical value to the patient (p. 11).

One solution might be to harvest cells at earlier stages and use these as a resource for later treatment testing, if the patients do not respond to the traditional treatment. However, cancers develop and change, and by the time the organoid "avatar" is ready for testing, the disease may have become something else than what was tested for: "During this time, patients being treated with first-line therapies may also develop resistance which may not be reflected in PDOs derived at a treatment-naïve stage" (Bose et al. 2021, p. 1020). Researchers also face the challenge that patients enrolled in experimental (Phase I and II) organoid trials often have a limited time window for treatment allocation. As one of our informants states, they must

> ...make sure we do not leave patients without treatment for too long. We actually do biopsy at the beginning of the last line of treatment, which is pretty short in colorectal cancer, mostly they have a scanner after two months and 50% of the patients have progressed after two months, so that give us eight weeks from the biopsy to be able to produce a report for the clinician to orient the treatment if the patients have progressed... What is difficult is that, once we get the report, because it is a Phase I-II trial, the patient may have died already. (senior researcher, academic)

Another of our interviewees says: "It shouldn't take 6 weeks, it is too long for a cancer patient. It should be done rapidly, it should be done by a standard educated technician, a trained technician should be able to run the assay". (senior researcher, industry)

There is some optimism among our interviewees that this can be done, although it is challenging:

> I think the gold standard for the return of clinical sequencing is around two weeks. I think that it is doable, but it is very difficult even in a very trained team - doable but you really need to train your people accordingly: the analyst multiplying the tumour cells or organoids, the radiologist, etc. (junior researcher, industry)

4.2.5 Gaining the Trust of Clinicians and Evidence-Based Medicine as an Obstacle

For researchers who are convinced that organoid models work, the main difficulty towards clinical application is to gain trust from the medical community. We have an issue here at the intersection of evidence-based medicine and trust/authority in the medical community. Two quotes from our epistemological hotspots interviews illustrate this:

We have targeted therapy which is very successful: if the patient has a particular mutation, it often translates to a particular drug. A lot of oncologists will depend on that in prescribing drugs for patients, but our data does suggest that sometimes it doesn't fall into that path, and if you try to convince an oncologist to 'maybe try something else', there's always a blockade... You have to understand that molecular targeted therapy came out of clinical trials, so it has been endorsed, in a way. That's why, from the very beginning, I said all organoids need is a clinical trial, so again, it has to be endorsed. At this moment, we are in a kind of embarrassing situation, where we definitely know that, organoids have a value, but then at the same time oncologists know that molecular targeting does have a role and trying to convince them to change to something else, it's not always easy. (senior researcher, academic)

And...

It turns out that the studies so far agree that the predictive value is very high, in the order of 80 to 90 percent correct predictions that the patient will or will not respond to a specific drug or drug combination. The challenge is that there are many guidelines, and oncologists cannot just ignore a guideline for a treatment of, say, a breast cancer patient at a particular stage of disease and say: 'well, I don't like the guidelines so I will go with the results of an organoid test'. There we need – and there are now companies that are doing this – to validate this very extensively before we can change the way patients are treated. (senior researcher, industry)

As the latter quote illustrates, the challenge is not only to gain the trust of clinicians but also to achieve sufficient documentation to change regulatory guidelines for preclinical and clinical research.

4.2.6 Biobanks and the Stratified Approach as a Solution to Technical and Practical Issues

So far, we have focused on factors that may hamper the patient-specific approach that we described as the first type of clinical delivery in our vision analysis. However, the stratified approach and use of biobanks has been presented—and will here be considered—as a solution to at least some of the technical and practical issues we have mentioned above. Francies and Garnett stated in 2015 that,

Finally, establishing and drug testing organoids from each patient would be a mammoth and expensive undertaking. We believe a more plausible scenario for the foreseeable future is the establishment of a biobank of thousands of organoids, encompassing the molecular diversity of the majority of tumor types. Genomic characterization and drug screening of the biobank would enable elucidation of molecular biomarkers that predict patient drug sensitivity and resistance.

In general, the patient-specific approach is seen as preferable, but stratified approaches may be preferable if there is not enough material, time or resources to grow organoids from the same person, or if the patient's tumour is inaccessible, as e.g., brain tumours often would be. It may also avoid unnecessary pain from new biopsies in patients.

The procedures at biobanks—and between biobanks—may be easier to standardise than between all different labs and clinics, reducing the problem of technical variability as discussed above. Biobanks may have higher success rates than one achieves in labs generally: "the most successful biobanking efforts report establishment rates of 70–95%, which can decrease in other settings" (Bose et al. 2022, p. 2)). Biobanks also have the advantage of being able to generate models similar to patients over time without being in a hurry and being limited to the same degree by contamination. Additionally, biobanks may be seen as "precious resources" when "patients have received neoadjuvant therapy prior to surgical resection, which may affect its role as prognosis tool in preclinical and clinical practice" (Yan et al. 2022, p. 11).

However, biobank establishment is not without problems either. As some of the patient groups that one wants to help in PM are small (or only one person!), "obtaining enough cancer samples is difficult due to the limited number of patients" (Ma et al. 2021, p. 7). One of our interviewees notes:

> In the future, you could imagine that you need your specific, your own organoid, or there would be this large bank and you could just say: 'within this bank where we have tumors that are like you…' (…) But we just don't have that (…) You need to have the diversity to reflect the population. (senior researcher, scientist)

We are here closing in on the fundamental problem of biobanks: It is not, to the same degree as the "patient-specific" approach, an individualised strategy, capturing inter-patient heterogeneity to the same extent. With the stratified approach, the vision is shifting to something different. It may be more practical, and less costly, but at the expense of personalisation itself. Whether this is a better approach, thus also hinges on the uncertainty about how heterogeneous the disease itself is. That is, uncertainty about the extent to which accounting for variation between individual patients is more or less important than addressing the other uncertainties about lacking standardisation or practical limitations.

4.3 Uncertainty Regarding Main Expectation C: Amenability and Documentation

Now we move on to examine the third main expectation: If the utility of organoids is there, it should be amenable to scientific documentation, or—at least—forms of evidence production that are acceptable to clinicians, healthcare authorities, insurance companies and other stakeholders. This includes two related questions: Can the utility be documented in a generalisable way? How can the clinician know beforehand that the diagnostic test (or rather the treatment it indicates) will work in the individual? To answer these questions, we begin by evaluating the clinically relevant evidence published up until our search date, and then the registered trials up until that time.

4.3.1 Summary of Published Clinically Relevant Evidence

As part of our evaluation, we will now evaluate the evidence status of the PDTO for PO vision in terms of published, clinically relevant studies. We take a wider look at the evidence than what is commonly done in health technology assessments to get a view of where the vision is in terms of realisation and plausibility. We are here first investigating clinical validity of the models as predictors, but most of all, the clinical utility (evidence on effectiveness in a complex clinical and social setting). The general claim in the vision is that organoid response is predictive of patient response. What kind of evidence is brought to support this claim? In which diseases? For how many patients and with which conclusion?

The following can be considered as an informed and critical review and synthesis on the material collected. All the studies included here have involved patients in one way or another. Whether or not the organoid has an implication for patient management is another issue because it depends on the type of study. As stated in the methods chapter, there were 55 publications on organoids in our material of published empirically relevant research. Cancer is the focus of 47 of 55 publications. The first clinically relevant publication on cancer is from 2018, and since then there was a steady increase by year, reaching 15 publications in 2021 and 9 in 2022 up until our latest search.

The USA had published the biggest share of clinically relevant empirical work, with Europe and China being other major contributors. China seemed to be gaining prominence with 11 of 12 publications appearing in 2020–2022, as compared to 9 in the US. This stands in contrast to the review material, where the US and Europe dominate with a substantially bigger part of the publications. One can therefore say that while China is taking a leading role in fulfilling the vision, the vision has been mostly generated in the USA and Europe. Other countries contributing to empirical publications include South Korea, Taiwan, Australia and Canada.

The studies that have been published so far essentially pertain to the patient-specific approach. No studies look at comparisons between the patient and an organoid coming from another, similar patient in a biobank (the stratified approach). We speculate that this is because the biobanks that are central to this approach are not yet comprehensive enough.

Of the 47 cancer publications, 21 studied metastatic cancer, at least in part. Many subtypes of cancer have been investigated. A further categorisation of the 47 publications on cancer is illustrated in Fig. 4.1.

Comparing the numbers of diseases covered in published material with the Fig. 3.2 in the vision analysis above, there is also here a striking dominance of GI cancers, with almost half of the cancer publications. What is also striking is the relative disappearance of prostate cancer and urological cancers in the actual empirical work. This may be because of difficulties with growing these organoids.

The empirical studies in our material are mostly about drugs, but 8 of the 47 cancer publications have radiotherapy as at least one element of the intervention, and 2 of them focus on radiotherapy only.

4.3 Uncertainty Regarding Main Expectation C: Amenability ... 71

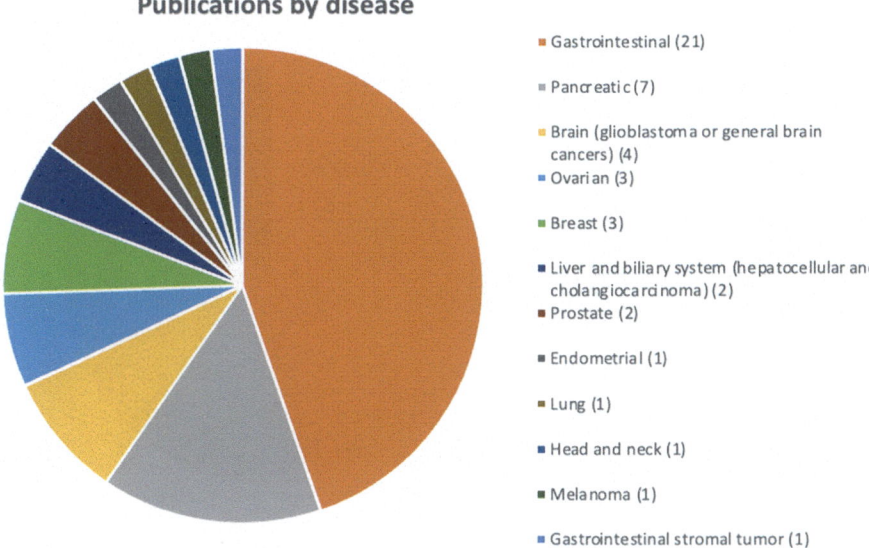

Fig. 4.1 The distribution of cancer types (by organ system) focused on in the published clinically relevant material (sum 47 publications). The picture is dominated by colorectal cancer (which accounts for most publications in the gastrointestinal category), but esophageal or gastric cancer is part of the 21 publications on GI cancers. The brain category includes glioblastoma and publications on "general" brain cancers. The liver and biliary system includes hepatocellular and cholangiocarcinoma. Gastrointestinal stromal tumour (GIST) is categorised alone as it is the only non-epithelial among GI cancers here

The number of patients included differs from case studies with only one patient to larger observational cohorts. Importantly, the number of organoids grown, and correlations established between patients and their organoids may be more representative of the information provided by a given study than the number of patients included. A total of 543 cancer patients having their clinical outcomes correlated with organoid findings have been reported in our material of published literature on cancer.[5]

The mean is 11,5 (correlations per study) and the median is 5. The pie chart below shows that a majority of the studies (31 out of 47) a majority of study reports 10 patient cases or less. 10 of 47 publications report on only one patient (Fig. 4.2).

Considering that these cases are distributed in many cancer subtypes and that the studies follow different methodologies, this number is relatively low. The general assumption that the proof-of-concept level has already been reached is therefore worth discussing.

---

[5] Considering the totality of our material, that is, including non-cancer studies, this number amounts to 908. It is notable that all non-cancer studies are on cystic fibrosis.

Fig. 4.2 Distribution of the number of PDOs/patients correlations established by study. The number (n) is the number of patients that have at least one organoid established. Not all patients enrolled have a PDTO. The n is the number of patients whose clinical outcomes have been correlated with their PDTOs

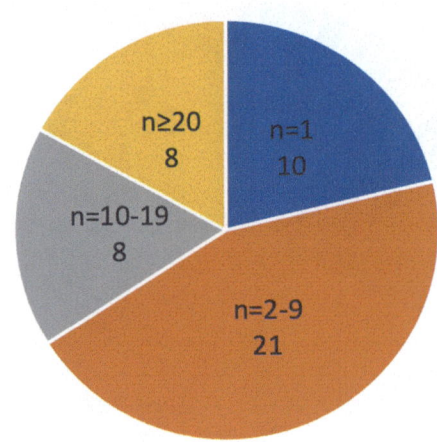

4.3.2 Discussion and Conclusion: The Evidence Status of the Vision

Looking at this above described, published evidence, what is the evidence status of the "PDTO for PO" vision? How far has the field got in documenting that the strategy works? In our vision analysis, we described that the field envisions a stepwise strategy in documenting their models for PM through two broad categories of studies:

First, **observational studies** (including parallel or co-clinical trials) that may be both retrospective and prospective and that aim to document the clinical validity of the models, their ability to predict what will happen in patients. This may be seen as "proof-of-concept" studies.

Second, **interventional trials**, which include prospective validation trials (cohort studies), RCTs, and also interventional single-case experimental design studies (or SCEDs). These test a situation where the model is used directly to guide treatment. Their aim is to document clinical utility.

The material is dominated by parallel and co-clinical trials that compare the organoids and the patient from whom they were derived when they receive the same treatment.

Only 8 of the trials on cancer are interventional. Being an "interventional" publication here does not mean purely interventional. Several of these publications are observational with one of a few patients having treatment selected based on organoid findings, and it is often unclear why just these patients are chosen and reported on among all the others. Only 2 of 8 studies are "purely" interventional (the key Ooft et al. 2021 study as well as Meier et al. 2022).

In the 8 interventional publications we have located, the sample sizes are generally very low with a total of only 29 cancer patients having been studied in this way. There

4.3 Uncertainty Regarding Main Expectation C: Amenability ...

are no RCTs to document the scheme at the time of writing, and neither are there Phase II or III trials more generally.

19 of the 29 cancer patients were from the single study by Ooft et al. (2021). This study stands out in our material in terms of methodological rigour and scientific weight, but it is still only a Phase I trial with no control group. Notably, this largest and most rigorously performed study, had a negative or inconclusive result, and it also illustrates practical feasibility issues in making the strategy work. This is sobering.

5 of the 8 interventional publications were single-case design studies (n = 1) (Loong et al. 2020 on glioblastoma, Guillen et al. 2022 on breast cancer, Reed et al. 2021 on brain cancer, Meier et al 2022 on hepatocellular cancer, Wu et al 2022 on pancreatic cancer). Two studies had 2–3 patients (Chen et al. 2022 on brain cancers and Narashiman et al. 2020 on colorectal cancer).

In sum, the evidence from the 8 publications with interventional studies documenting clinical utility must be regarded as weak by EBM standards. Clinical utility of the models can generally not be said to have been shown. Importantly, cost-effectiveness has not been studied at all. It is thus still a very open question what these models can and whether they will be used in clinical care.

But what about observational evidence supporting clinical validity and the concept that these models are generally predictive? The observational studies in our material generally claim correlations in responses between patient clinical responses and their PDTOs. It is therefore tempting to conclude that results are very promising, that the predictive capacities of the models have been documented, and that we have a "proof of concept". Importantly, this also seems to be the general consensus in the field. In the interviews of our "epistemological hotspots" material, we find a near total agreement that observational "parallel" studies is already an achieved level in the documentation process, and that what is now lacking, is mainly evidence from interventional trials investigating clinical utility. As one interviewee states: "That kind of parallel trial, we have done, we know, the literature is packed with it that" (senior researcher, academic). Another interviewee:

> Now what we need is a demonstration... what we question here is the clinical utility, even though the validity is demonstrated and actually they are good avatars of the tumor ex vivo, that doesn't mean that this can improve patient survival overall, or relapse-free and so on. (senior researcher, academic)[6]

Is this a valid conclusion? As a critique, we must first note that there are many small, retrospective trials. And it is often unclear why precisely the patient or patients reported on are selected from a larger population. Are they reported on because they were the patients in whom the strategy seemed to work or seemed most likely to work (i.e., a form of selection bias or reporting bias where results are skewed because certain data are reported and others not)?

---

[6] See also this quote: "We see oftentimes that if a patient responds to a particular drug, the organoid does as well. That type of correlation, I do not think it is a problem to generate, and I think numerous groups are doing that, and building the case for convincing the regulatory people, the FDA in the US and their counterparts in Europe, that we should try to move this clinically. But, and maybe this is what you are getting at, there does need some better or more formal sort of documentation system as this moves clinically, and I do not know that that exists" (senior researcher, academic).

Notably, the predictive capacity of the models has been studied only for the patient-specific approach. The studies thus only illuminate a small portion of the cancer disease panorama. Major cancers like prostate cancer are largely missing. Diseases where fundamental problems with the models or practical issues are more limiting are likely not the first to be investigated.

Moreover, it is—as always in medicine—an early phase of research, where the circumstances may not reflect real-life conditions. As one interviewee admits,

> ...so far it looks extremely good – but, to be honest, of course publications in Phase I and II trials tend to be more positive than what happens in the real world, because everything is much better controlled, patients are better taken care of, I assume. (senior researcher, industry)

However, the most important concerns that challenge the "proof of concept" conclusion are publication bias and lack of standardisation in the field, particularly of developmental processes, which render results non-generalisable.

Publication bias is the problem that some studies, mainly studies that do not show a positive result, are never published. Publication bias creates a false impression of the validity of existing findings. This is obviously a problem in many fields of science and medicine, not only organoids (de Vries et al. 2018). Nonetheless, we find reason to suspect that it is a significant problem here.

Firstly, in our evaluation of the registered trials (see below), 16 observational studies were listed as either completed or of unknown status, but we found only one of these that had results published (in a conference abstract). We have emailed investigators of the registered trials but have received no replies at the time of writing.

Secondly, our informants in the "epistemological hotspots" voiced clear concerns that publication bias is substantial in the field:

> In general, whenever something fails, you are not going to publish it because it is extremely difficult to publish. We actually did an effort to do that, and it took ages, it was negative data but because we were the first we thought: 'well, it's relevant because people are trying this anyway'. I think somehow you have to also definitely document the negative results, and I know lots of negative results have arisen in the past years, but people don't bother, they just go to the next project and try to find something positive. (junior researcher)

Interviewee 7:

> Unfortunately, there is not a high impact avenue to write a paper that says 'hey, we tried this, and it doesn't work!' Who wants to read that paper, right? (senior researcher, academic)

Another interviewee states that,

> You see the correlation, but they only publish the results so that they found the correlation, and for the unpublished, uncorrelated results they say, 'what did we do wrong? what did we do wrong?' and they don't publish that. For every published article of correlation, I am sure there are hundreds that show no correlation. (senior researcher, industry)

Although this last statement (about the hundred studies) should perhaps not be interpreted literally, there is likely substantial publication bias in the field.

These problems, and especially publication bias, may undermine trust in the models. In conclusion, while the results from observational trials certainly do not

4.3 Uncertainty Regarding Main Expectation C: Amenability … 75

negate the overarching promise of better predictions and utility through personalisation, we cannot fully agree with the conclusion that the proof-of-concept phase has passed, and that the clinical validity of the models has been established. At the same time, it is easy to agree with interviewees at "epistemological hotspots" who argue for more robust clinical studies as the way forward:

> I think anything you do right now with organoids in the clinic should be documented in a clinical study or clinical trial, because it forces you to adhere to good clinical practice and good lab practice. I think we had too many studies that were retrospective and just looked at... you know... after they collect some samples to see if there is any correlation (…). I think it is nice as a proof of concept, very early phase development of organoids, right now we reach the stage that anything you try should be in a clinical study, which forces you to adhere to certain standards, quality standards. (junior researcher, industry)

This brings us to our evaluation of the material of registered trials.

4.3.3 Summary of Registered Trials: The Translational Pipeline

The consensus in the field, and we concur, is that more rigorous clinical trials are needed to show clinical utility and cost-effectiveness, which has not been looked at so far. We have also argued, contrary to voices in the field, that there is a further need to document clinical validity and predictive capacity of models, especially for diseases that have not been investigated so far. To further evaluate the plausibility of the vision and the process through which it is realised, we now move on to evaluate what evidence may be forthcoming and that is visible in the "translational pipeline" of registered trials.

Our analysis of our registered trials material shows that the majority of the registered trials aim to explore the utility of organoids for patient-specific drug screening, and only a few (3 trials) explicitly highlight the aim to develop biobanks for future use in stratified risk profiling and treatment screening. All registered interventional trials use a patient-spec a patient-specific approach. In sum, that means that what is on the translational horizon is almost purely about the patient-specific approach.

In terms of diseases being studied in upcoming research most registered trials focus on cancer, and among the cancer trials, gastrointestinal cancers dominate (34%) (see Fig. 4.3).

Breast cancer, lung cancer and pancreatic cancer also feature prominently in the distribution, while prostate cancer and urological cancers—as in the published literature—are still missing, despite their prominent place in the visionary literature. This may again be due to problems growing them. In our vision evaluation above, we described a two-tiered vision for documenting that the organoid approach works, with different types of observational trials being followed by interventional ones, which we also explained in some detail. Based on the study description for the registered studies, we categorised the 96 cancer trials we found into observational trials,

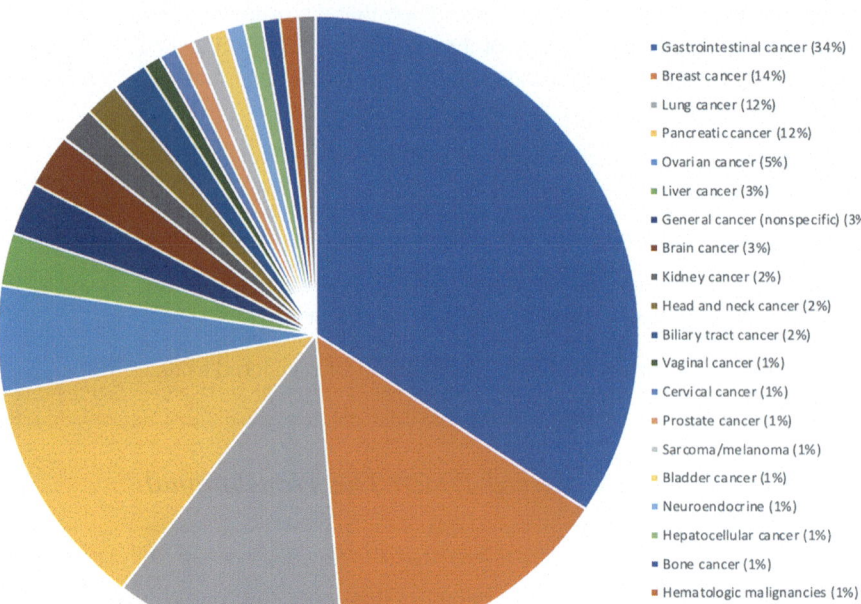

Fig. 4.3 The distribution of cancer types among the registered trials focused on cancer

interventional Phase I-III trials, as well as a category of interventional trials that we call "experimental/unclear" as they were hard to place (see Fig. 4.4).

Observational studies to validate models are often seen as a precondition for implementing organoids as tools for clinical decision-making in interventional trials, and the high number (relative to interventional trials) is therefore not surprising at

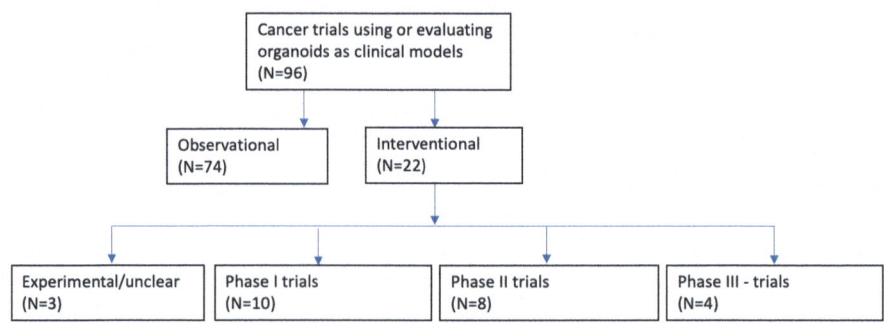

Fig. 4.4 Distribution of study types among the total number of registered trials on cancer. Three studies have a combination of Phase I and II trials and are counted twice in the table, explaining why the number of specified trial type at the bottom exceeds the number of registered trials in total

4.3 Uncertainty Regarding Main Expectation C: Amenability ...

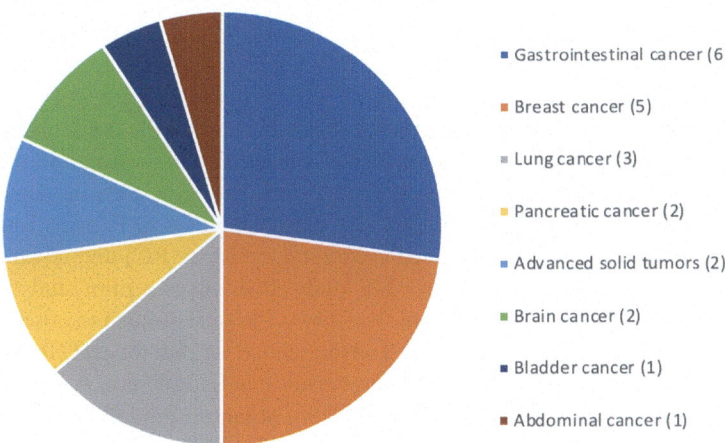

Fig. 4.5 Distribution of different cancer types among the interventional trials

this stage of technology development. At the same time, compared to the published material, we see a somewhat increased focus on interventional studies. This is to be expected as the field perceives that the observational "proof-of-concept" level has been attained.

The number of patients in the observational trials varies from 33 to 2000, but most trials planned to recruit between 50 and 150 patients. The focus of the larger trials is typically pancreatic cancer, colorectal cancer, breast cancer, and general cancer. The company Known Medicine Inc.[7] (Utah, USA) was recruiting up to 2000 patients in the largest planned trial (NCT05338073), with different types of cancer. The purpose was to determine the predictive power of the company's organoid-based test, Known 3Dx Test, for patient-specific testing of treatment response. Importantly, the size of this, and similar trials sponsored by medical companies, points to a possible future where the largest biobanks and best-validated models may be offered by commercial providers.

We found 22 planned or ongoing interventional trials. The types of cancer studied in them are shown below in Fig. 4.5.

See Appendix B for a more detailed overview of the interventional trials.

Phase I trials commonly refer to "first-in-human" trials, testing of the safety of different drug doses of new treatments, but can also refer to trials involving patients with advanced diseases that do not respond to standard therapy, e.g., end-stage cancer patients. The 10 trials we found here are of the latter kind, where patient-specific drug screening using tumour organoids are used to guide treatment selection, e.g., by comparing the response rate of different anticancer in one work for consistence in the document. The diseases studied in these Phase I trials include breast, brain,

---

[7] https://www.knownmed.com/.

gastric, lung, bladder and advanced solid cancers. They typically have a low patient number and no control group, several have only 10 patients. The largest Phase I trial, and also the only one published, is Ooft et al. (2021), discussed above. Several of them will not be completed until 2025.

Phase II trials are considered stronger evidence than Phase I trials, because they typically involve more participants, higher control over variables and randomisation. In the 11 Phase II trials, the patient number is generally higher than in the Phase I trials, although two trials recruit only 10 patients. The largest Phase II trial on the list (the ORGANOTREAT trial, NCT05267912, conducted in France) is expecting to enrol 1919 patients with advanced colorectal cancer (CRC) and advanced solid cancers. It thus stands out as a very big study. It involves a pilot-study restricted to advanced colorectal cancer (Phase I) and two Phase II studies in advanced solid cancers, where one is single-arm (2A), and the second is randomised (2B) to compare the efficacy of organoid-guided treatment versus standard of care. A cross-over will allow patients in the control arm to benefit from organoid-guided treatments.

The 8 registered Phase II trials focus on cancer: breast, gastric, bladder, colorectal cancer, lung, and abdominal cancer, as well as different advanced solid cancers. 4 of the 8 have randomised controlled designs. Most have an expected completion date in 2025 or 2027, so it will take several years before the results of these trials will be published.

Phase III trials are a phase of research that involves more participants to study the safety and effectiveness of a treatment, compared to a control group receiving no or standard treatment. Four registered trials on organoids currently qualify as Phase III trials studying the use of patient-specific organoids to guide treatment in individuals. These studies will recruit 120, 54, 93, and 200 patients and are focused on breast, gastric, colorectal, and pancreatic cancer, respectively. All except one are conducted in China, supporting the notion that while the USA and Europe have been central in the origins of the vision, its realisation is substantially taking place in China. One was located in Finland. The motivation of this study is to "reduce overtreatment of those that most likely do not benefit from additional treatment" by identifying non-responding patients. The small patient samples in many of the Phase III trials (see appendix B), compared to traditional Phase III trials, should be noted. Expected trial completion dates for the Phase III trials are 2023, 2023, 2025, and 2031.

4.3.4 Discussion and Conclusion: Registered Trials and the "Translational Pipeline"

In conclusion, regarding the "translational pipeline", we can see more research appearing in the coming years. In absolute numbers, more interventional trials will be conducted, and there is reason to believe that what is now in the pipeline will generally be more rigorous in terms of methodology than what has already been published (which has often not been registered). At the same time, observational

4.3 Uncertainty Regarding Main Expectation C: Amenability ...   79

trials still dominate the picture. This will be needed to more rigorously document the predictive capacity of the models in more diseases.

The first Phase II trials are appearing, all but one with randomised designs. Importantly, 6 of 8 registered Phase II trials are randomised in design, and there is one large Phase II trial (ORGANOTREAT conducted by Jaulin and colleagues). On our list of interventional trials, this appears as the most remarkable trial. This also raises questions about what counts as strong evidence in this context, e.g., whether there will be the same emphasis on randomisation and large patient numbers in future trial designs.

Still, we do not know how many of the registered trials will end up being published, and there is still a danger of publication bias. We searched for published results following trials registered as completed trials and found no published results on organoid models on the clinical trial databases or as full articles in international journals. One trial (NCT03544047) was referred to in a poster published in the *Journal of Clinical Oncology* by Acklin-Wehnert et al. (2023). We emailed the PIs of completed trials to inquire about publication of results we may have overlooked or forthcoming publications. Publications are currently under review following trial number NCT04342286, but we did not hear back from others.

Importantly, no planned or ongoing interventional trials are about the stratified approach, although at least three of the observational trials seem to be. The vast majority of trials on cancer are aiming for patient-specific screening, showing that the part of the vision using patient-specific models is dominant in ongoing trials. This also underscores the importance of discussing the benefits and challenges of this approach. In the registered trials, biobanking was mentioned explicitly as a primary or secondary aim in only 10 of the 109 trials. However, among these were some of the largest trials with several hundred patients, and two above 1000 patients. A few trials aim for development of a "living biobank" for stratified screening approaches via matching of patients and organoids from patients with the same tumour types (for examples, see NCT03655015; NCT03896958). This approach can possibly address the translational to develop sufficient organoid material challenges of developing sufficient organoid material for "real-time" drug screening, although at the trade-off of not being patient specific.

In sum, we find a "translational pipeline" that seems somewhat meagre, but with more evidence forthcoming. However, it seems clear that it will take at least 10 years before a substantial evidence base is produced. What is seen in the pipeline may therefore quite slowly document the patient-specific approach as useful (if that is what it turns out to be), and even more slowly the stratified approach. The plausibility of the vision therefore still rests substantially on the theoretical considerations we have discussed above (main expectations A and B).

We now move on to discuss two main areas that we see as epistemological hurdles to process and overcome for the field to be amenable to documentation. The first is the need for standardisation on different levels.

4.3.5 Hurdles: "Standardisation, Standardisation, Standardisation"

Having looked at the published clinically relevant evidence, and the registered "translational pipeline", we now turn to discussing obstacles in documenting the PDTO for PO scheme. As we saw in Sect. 4.2, lack of standardisation creates variability between models and technical noise. However, there is lack of standardisation in the field on several levels, which need to be addressed to improve its evidence basis. This is summed up candidly by one of our interviewees at the epistemological hotspots:

> Where do we need to go? Things like success rate of cultures, simple logistic things, how we can make these cultures, quick readouts... That is important for cancer but not so important for all the types of applications. We still need to learn for specific contexts of uses, how we can use these types of models. I think for all these types of uses it is about standardization, and standardization, and again standardization. Because most of it is still in the academic domain and we screw around, we don't have good quality control measures. (senior researcher, academic)

At the most basic level there is a need for standardisation of the organoid concept itself. Some entities that are called organoids, are not organoids, and perhaps vice versa. This will create linguistic uncertainty about what the evidence base is about. This is acknowledged in the field: "First, the definition of organoids needs to be standardized. Some articles use "organoids," but do not meet the definition of the organoid... Notably, certain researchers have confused the concepts of tumour spheroids and organoids" (Luo et al. 2022).

Furthermore, the "PDTO for PO" vision promises "functional" precision medicine and "molecular agnosticism". One may bypass the need to understand what mechanistically goes on in the model: If the model predicts, it predicts. Statements from our epistemological hotspot interviewees illustrate this idea:

> We don't need to understand everything about the tumor to be able to provide the right treatment. This kind of objective information, which is a response to a drug that you can obtain, is very important and can be used to the benefit of patients and as a decision-making tool for clinicians. (senior researcher, academic)

And...

> I think it will definitely get its place in clinical decision making, especially in which you have different structure and signals, you have like multiple drugs at the same time, and also which you know you are not very much aware yet of the mechanism of work of the drugs, so you can just blindly throw the drug on there, you do not have to know how it works, you just have to observe whether your tumor cells are dying or not. (junior researcher, industry)

This feature of organoid models, that they may predict even though one does not have full knowledge about how the model works, is similar to machine learning models that also to a certain extent "model themselves". As Walker and colleagues (2019) point out, this also creates a form of "black box problem", similar to that of artificial intelligence. This may pose problems regarding transparency, standardisation and trust in the models. The fact that the "biomarker" in "molecularly agnostic"

4.3 Uncertainty Regarding Main Expectation C: Amenability ...

organoids may not be a specific molecule or genetic sequence, but a more complex biological readout that is dynamic and not static, poses another problem: How will the field define what to look for in the models that is supposed to predict that a person will have an effect from a certain treatment (Letai et al. 2022)? Veninga & Voest (2021) put this bluntly in a review:

> While heterogeneous drug responses may recapitulate the unique biology of each tumor specimen, clinical translation needs robust readouts with predefined thresholds for drug sensitivity. Despite interesting correlative findings, a key limitation of these studies is the absence of such a validated organoid-based decision model on the basis of which a patient will or will not be treated. (p. 1192)

How does one define and standardise the diagnostic thresholds? For the research on organoids as predictive biomarkers to become more reliable, generalisable and actually report on the same thing, they also need to use the same readouts and thresholds. This challenge is exacerbated by the limited mechanistic knowledge about what goes on in the organoids.

In defining the "functional biomarker", one proposed method is to use systems biological, computational tools such as machine learning. Brazovskaja et al. (2019), for example, state that, "As the field progresses, robust computational strategies will be required to integrate the data and make biological sense of what is sure to be high-information content and extremely complex data". Such tools may be a way forward but are also currently challenged by limited standardisation. A related problem is the definition of what to look for in the patients that will be correlated to the organoid findings. Mackenzie et al (2022) concluded: "a lack of standardised quality control and drug treatment endpoint analysis methods currently prevent integration into the clinical space".

Additionally, standardisation of what drug concentrations in organoid are supposed to correspond to the drug concentrations that are effective in patients is a problem. This sentiment is also faced at our epistemological hotspots:

> What is difficult for instance is the pharmaceutical kinetics of a drug. How much drug do you need to add to an organoid to make sure that you have a similar amount of drug as compared to in vivo conditions? (senior researcher, academic)

And…

> …we worked a lot with pharmacists to be able to find relevant concentration of the drug test on the organoid as compared to what the patient would receive, because you need to make sense… of course any drug would kill the organoid ex vivo, that doesn't mean you will be able to give the drug to the patient at this concentration. So, you need to start from the concentration you are going to give to the patients if this drug is a hit in order to be able to set up the drug test at the right concentration (senior researcher, academic).

Moreover, there is a question of how many organoids per patient you need to test a drug for the results to be reliable. If one grows only one, the result may be borderline, or wrong by chance. This also creates a need for defining standards.

Having analysed the published clinically relevant evidence and the translational pipeline of registered trials, we have seen how results are in ways that are hard to compare. Sometimes the reporting also lacks important levels of information. This is a problem for generalisability of results and interpretation more generally. It creates a need for standardisation in reporting results: "Researchers should aim to adhere to methodological standards when reporting results, to facilitate study quality assessment (including potential biases) and study result interpretation" (Wensink et al. 2021, p. 30).

Finally, perhaps the biggest concern is related to the above-described lack of standardisation in the very making of organoids. In the materials used, and at several stages in that process, there is variation between clinics and research projects. This means that the results that are found in one study cannot be assumed to apply to any other clinic. In other words, there is lack of generalisability. As researchers in the field also acknowledge, "…results derived from PDO studies can often not be compared across publications because of the variability in microenvironmental factors" (Bose et al. 2021, p. 1021), and "drug screening results would also be difficult to interpret due to institutional variations" (Foo et al. 2022, p. 12). This creates a need for replication of studies under standardised procedures.

At our epistemological hotspots, the need for standardisation is also voiced very clearly. At the same time, it is clear, as Luo et al. (2022) state, that "although there is no consensus on the standardization of organoids, some results have been achieved" (p. 14). Generating a consensus on the issue of standardisation is an important first step. However, this may not be easy, as we turn to below.

4.3.6 The Tension Between Standardisation and Personalisation

There is an important and fundamental tension between standardisation and personalisation (Green et al. 2022). Standardisation also has downsides.

At the most basic level, standardisation may be at odds with scientific and clinical creativity. "Sometimes standardization is important, but it can also be constraining on what you can learn if you can't play around with your culture conditions," noted Lancaster, a pioneer in the field.[8]

More specifically, standardisation of organoid culture conditions may come into conflict with the goal of personalisation itself. This realisation is stated very succinctly by Luo et al (2022):

> Many scientists believe that improving standardization is an important step to expand the clinical application of this technology. At present, organoids are mainly applied in personalized clinical medicine, and organoid culture is carried out for different patients to select the most suitable drug for treatment. Different patients' organoid culture requires different conditions. Currently, standardization may limit the application of organoids in personalized medicine. We think the standardization of organoids should be the standardization of different situations, not a generalization. (p. 14)

---

[8] https://www.stemcell.com/nature-research-roundtable-organoid-applications.

The quote points to a tension between the call for standardisation of protocols and laboratory techniques for growing organoids on one hand, and the need to adapt culturing conditions to also account for variation in the tumour microenvironment of individual patients on the other. The articles stress that organoid growth rates and results can depend on the origin of the 3D matrix (animal or human material), the composition of the medium, as well as the biomechanical properties of the matrix. Factors such as matrix stiffness and axial strain are increasingly recognised to influence cancer development, prognosis, and treatment responses (Gkretsi and Stylianopoulos 2018; Ishihara and Haga 2022), highlighting how the relevant patient variation may not be reducible to genetic composition or even the composition of growth media. Although the underlying idea of personalised medicine is that the individual is unique, the field must find a way to balance the need to account for variation while standardising some aspects to ensure practical feasibility and validity of the models. The quote by Luo et al. (2022) above suggests a possible middle way between universal standards and a more "stratified" approach with situation-dependent standards. But where to most meaningfully draw the line between standardisation and relevant variation is an open question—and one that is intertwined with ontological uncertainty about cancer itself (see also Green et al. 2022).

A similar issue arises in discussions about how to best calibrate PDO models to allow for upscaling or estimation of clinically relevant drug doses in patients:

> The (…) challenge is the calibration of PDO models for clinical decision-making. The ex vivo test results depend on drug concentration or radiation dose and treatment duration. (…) drug responses can be quite different due to inter-personal tumor heterogeneity. If we collect data at multiple drug concentrations across a range of tumors to identify the calibrated drug concentration of the PDO models, then we lose the key for personalized medicine. (Wang et al. 2022a, b, c, p. 6)

Wang et al. (2022a, b, c) stress in their article that PDO models could help "personalise" drug dosages via comparison of the effects of drug dosage or radiation intensity on "healthy" and "diseased" organoids from specific patients. However, since many additional factors influence drug metabolism and treatment response, there is a call for a more systematic evaluation and calibration of dosage scaling from PDO models to human patients. Yet, this can be done only if some standards are defined for comparison—which may come at the expense of the quest for personalisation.

Finally, the persistent challenge between standardisation and variation is also present at the level of diagnostic thresholds in precision medicine, where many diseases are split into disease subgroups, e.g., based on genetic markers. This also means that the patient groups multiple in number but decrease in size. In cases where there are one of a very few people with a certain disease, the concept of standardising what is a significant readout may become impossible using statistical methods, leaving plenty of room for clinical judgement (see below). To address the problem of decreasing patient populations, personalised medicine may paradoxically have to rely on population-wide studies, where as much data from as many patients as possible can be combined to reach statistically significant numbers for the specific patient subgroups (Hoeyer 2019). For the organoid field, this would mean more concerted efforts to establish databases and biobanks providing knowledge resources for evidence on "similar patients". What the relevant level of variation is, however, is still an open question, and it is not uncommon to encounter statements in the literature that each patient is unique. This brings us to the vexing question of how to know and handle epistemological uncertainty in n-of-1 situations.

4.3.7 N-of-1 Situations: Knowing What Works in the Unique Case

The "PDTO for PO" vision is a novel part of the broader promise of changing evidence-based medicine so that treatments are no longer "one-size-fits-all." Still, we have seen how the organoid field still primarily seeks documentation through population-based studies. As outlined in our vision analysis, however, there will be situations in PM where such studies are not feasible to perform or seen as relevant. As stressed by powerful institutions like the US Federal Drug Administration, the question then becomes what evidence is needed (Woodcock and Marks 2019). An editorial in *New England Journal of Medicine* from 2019 asked: "In these "N-of-one" situations, what type of evidence is needed before exposing a human to a new drug?" (Woodcock and Marks 2019).

The issue of how one personalises in medicine, how one knows what is best for the individual, and the tension between EBM (as it has been practised) and personalisation have been ongoing for years, but have taken a new turn when prestigious biomedicine and cutting-edge biotechnology has adopted the PM concept and new treatments are sought marketed (Beckmann and Lew 2016). These debates may for example be about reliance on case histories, physiological and broader biopsychosocial knowledge for personalised decision-making as well as clinical judgement and practical wisdom (Tonelli and Shirts 2017). This also has led to a renewed interest in single-case design studies (SCDs) for reducing (epistemological) uncertainty and know what works in the individual (Nikles et al. 2021).

Organoids can be seen as offering an intriguing new layer of information in single-case design studies. Previously, clinicians had two levels of information.

(1) *Before providing treatments*, clinicians have relied on whatever relevant population-based evidence there is, previous case histories and clinical experience, as well as physiological, molecular and genetic knowledge when deciding to try a treatment in a single case.
(2) Then *after the treatment is tried*, new information arises in the patient indicating effect or not.

Organoids becomes another level of information on top of (1), and before (2). It is a form of test, aiming to predict what will work in the individual, but it is something more than the genome or molecular information. It aims to be a *patient-specific* "avatar" of the tumour itself, the results in which will be directly relevant for patient outcomes. It is a form of n-of-1 experiment in something that ideally would be a

4.3 Uncertainty Regarding Main Expectation C: Amenability ...   85

perfect representation of the person's biology. Or, if many organoids are generated from the same patient, it is a n-of-many experiment of high relevance to one person.

A main promise in the "PDTO for PO" vision is that organoids will decrease the uncertainty in making decisions about treatment in the individual case. However, this rests on the assumption that what is observed in the organoid is sufficiently predictive in individual cases to be a basis for decision-making. Where there are no other cases to compare the individual to, and no group-based trials are possible, that leaves us with an evidence conundrum. How can we trust the organoids when there is no other evidence pertaining to the individual case?

The antibiogram analogy that also came up during our interviews (see above) is interesting here: The readout in organoids is compared to the readout that microbiologists get when they test antibiotics on bacterial cultures in a dish. Medicine has trusted these readouts for prediction of resistance and responsiveness to antibiotics for decades: How do they know these "personalised avatars" in a dish of the person's infection are predictive of the person's problem? Why do we trust them as predictive and useful? Future research should look at the processes microbiologists have gone through to validate these models. This should be of high relevance to the organoid field.

Walker et al (2019) have argued that organoids involve a new and different form of inference in personalised medicine, "which could enable it to provide predictions about treatment responses that are truly personalised" (p. 117). They argue that organoids will rely on a form of correlational knowledge. They write that...

> [T]he correlational data are used differently in this approach than they are in other black boxes—including traditional correlational studies and other PM strategies. In those approaches, predictions about how an individual patient is likely to respond to treatments involve inferences drawn from other people's responses. In particular trials, those other people may be more or less similar to the individual patient. PM strategies enable more personalised treatment by relying on new ways of identifying or defining relevant similarities, which improve the strength and reliability of the inference from others' treatment responses to the individual patient's treatment response. (Walker et al. 2019, p. 117)

Walker and colleagues here express the idea that PM will allow researchers to more accurately identify clinically relevant similarities between patients, thus making inferences between patients more reliable. Others have similarly emphasised that intervening on organoids from the same person allows for more direct inferences of treatment response in the individual patients. For example, philosopher Giovanni Boniolo has claimed that patient-derived organoids enable a new "science of the individual":

> Since the Aristotelian discussion of the architectonic of knowledge, it has been accepted almost as a platitude that there is no science of the individual. [...] Whenever you study the primary cancer cells of a given patient, you are also studying tumor heterogeneity, that is, something at the universal level. But you are also studying the particular disease of that particular patient, that is, you are also studying the individual. [...] Put in a different way, within the field of tumor heterogeneity we have the possibility of doing science of the individual, since the tumor cancer cell population actually is an individual (patient) *in vitro*. (Boniolo 2017, p. 29)

Yet, as discussed in previous section, organoids do not straightforwardly represent the individual patients and are still in need of validation as a tool for clinical decision-making (see also Green et al. 2022 for a discussion).

One initial thought in the organoids field is that—even in the absence of interventional studies on a specific combination of cancer and treatment—one may rely on the general assumption that organoid models are predictive based on observational research that has shown their clinical validity. This is illustrated by one of our interviewees.

> If we can show success and correlation in different cases, if we can show that we have positive correlation or good outcomes in colorectal cancer, brain tumors, lung cancer, and now we get this rare adrenocortical carcinoma (…) then because we have proven our technology platform across other areas, I think we would have confidence that we probably can do it well in this other area. (senior researcher, academic)

This general trust in organoids as predictive models challenged by the problems of accuracy (representation) and variability as well as publication bias that we have discussed above, entailing considerable uncertainty. However, if such prior knowledge is coupled to single-case design studies in each case, this may also in some cases—e.g. where no other treatment is available, prognosis is poor and side effects and costs are acceptable—be an option.

One may also envision single-case design studies of various rigour pooled together in case series and databases. If these grow to a sufficiently high number, they provide a form of proof that something works, and also a degree of generalisability to new patients with the same characteristics (Schork 2015).

Several of the studies we have located in our material of clinically relevant evidence are SCDs. Such trials may have a high status in the evidence-based medicine hierarchy of methods—when performed in a rigorous fashion. However, other, less rigorous single-case strategies have also been envisioned (Schork 2015; Kane et al. 2021; Vogt 2025 in press).

5 of the 8 interventional publications in our material of published evidence were single-case design studies. None of these are rigorous trials with randomisation, repeated interventions, blinding or placebo controls. They can all be classified as pre-post trials or just case histories, that is, trials where one observes what happens in an individual after treatment (Kane et al. 2021).

Importantly, the other interventional cohort studies in our material may be understood as aggregates of SCDs. This is a general point: In a situation where patients get different treatments based on different results in their organoids reflecting inter-patient heterogeneity, all the participants in larger studies can be regarded as single cases (aggregates of $n = 1$). Coming cohort studies and RCTs can also be seen as aggregates of single-case design studies, where each of the cases have unique features and individualised treatment and must be evaluated individually within the larger trial.

This all illustrates a need to develop work on single-case design protocols in the organoid field, and standardisation of these protocols so that they can be aggregated into larger pools of cases that have common denominators. That organoids represent a new layer of information in single cases, also means that the possibilities they offer should become part of ongoing efforts to develop and formalise single-case design studies (see e.g. Woodcock and Marks 2019; Nikles et al. 2021).

In general, in PM and the organoid field one will in some cases need to develop evidence not for a specific disease-treatment combination, but for a more general strategy within which one accepts that considerable variation exists.

An option for documenting that therapies work in n-of-1 situations that we do not find mentioned in the literature, is cluster randomised trials (also called group randomised trials). Here, instead of randomising individuals to different arms of the study, one randomises different groups that are comparable to different arms (e.g., different clinics). In practice, in the absence of other evidence, one could for example randomise different cancer clinics into two arms, where one arm provides off-label cancer treatments based on organoids findings in rare cancers without other forms of evidence and compare survival here with survival in clinics with, e.g., standard treatments or genomically guided precision medicine.

The question of "how we know what will work for whom" when $n = 1$ is a vexing one in organoid research. In many ways it may increase uncertainty about what works, or, perhaps more precisely, increase the uncertainty about what epistemological standards one should adhere to in evidence-based medicine. N-of-1 situations take medicine back to the age-old problem of clinical judgement in the individual case and will involve handling uncertainty that cannot be reduced statistically but is qualitative in nature. In other words, it will bring the organoid field into contact with "the art of medicine" (Vogt and Hofmann 2022; Vogt 2025, in press).

4.4 Conclusion of Vision Evaluation

We now reach the conclusion of this assessment of the "PDTO for PO" vision. In our vision analysis we described its main, overarching promise as being able to *improve predictions of what the patient with cancer disease needs and provide significant clinical utility through personalization of treatment*. We have evaluated this overarching promise by dividing it into three main expectations, each with more concrete promises, and underlying assumptions, conditions and prerequisites:

- The first main expectation is that PDTOs can yield models that more accurately capture the true complexity of a patient's specific tumour as well as variation between individuals (inter-tumour heterogeneity), and that they therefore will give better predictions for the individual than existing guidelines.
- Second, that the proposed patient-specific and stratified strategies will be practically feasible.
- Third, that the strategies are amenable to methods that can document them and provide evidence that they will work for individuals.

4.4.1 Credibility and Hype

How credible or plausible is the overall promise of the vision?

In our vision analysis, we distinguished between "revolutionary" versions of the vision, promising revolutions or paradigmatic shifts in medicine—and more modest versions that paint a picture of incremental gains and specific applications.

All in all, our perception based on our vision assessment is that the organoid for PM vision is not particularly hyped, comparing it for example to the P4 medicine vision (of predictive, preventive, personalised and participatory based on computational models of personal data) that two of the authors have studied before (Green and Vogt 2016). The more modest versions of the vision seem to be reasonably calibrated against the potentials and limitations related to organoid models and the current clinically relevant evidence.

That said, the more revolutionary versions of the vision seem hyped. To state, for example, that organoids have the potential to "revolutionise future patient treatment" (Clark et al. 2022), is a long shot considering the limitations we—very much based on the perceptions of players in the field—have pointed out.

Firstly, we have pointed out several limitations that challenge the first main expectation of better representation of the individual patient. Most fundamentally, organoid technology has several significant problems in representing the specific characteristics and context of the patient's tumour, which is its most central underlying promise. Our main conclusion is thus that organoid technology does not currently fulfil its promise of accounting for context: It can do this to a certain, but still significantly limited degree.

This lack of accounting for context in turn is linked to a weakness with the main assumption and design principle of organoids: That they will self-organise and "model themselves" without great control of many parameters. Without an accurate, constraining environment, this self-organisation process in fact becomes more "disorganised". This has resulted in the realisation that more control of different parameters in development is needed, something that therefore becomes a main promise of organ-on-a-chip technology (Hofer and Lutolf 2021).

Despite its promises, we find that the organoid field clearly represents a form of biological reductionism, where physiology and personal disease are reduced to something that is supposed to be represented at the cellular level. We register with interest that several agents in the field acknowledge this, and point towards more "holistic" technologies for solutions (mainly OoC) (see e.g., Hofer and Lutolf 2021).

It is important to note that several of these fundamental issues are not only about technical limitations, but fundamental lack of knowledge: About the mechanisms mediating successful establishment of organoids, about biomechanical cues modulating development, about what an extracellular matrix should be like, for example. This, in sum creates considerable—but not quantifiable – *model uncertainty* about what these models can achieve.

4.4 Conclusion of Vision Evaluation   89

At the same time, we by no means deny that PDTOs can be sufficiently predictive because of their representational limitations. Although they are not—and will not—be perfect in mimicking in vivo counterparts, they may in some regards be superior to other models (both animal models and 2D cultures) and offer productive ways forward. This is why the more "modest" forms of the vision are more plausible.

The degree to which the vision is plausible also depends on what kind of disease we are considering. We find that gastrointestinal cancers are dominating, followed by e.g. lung, breast and pancreatic cancer, but common types of cancer such as prostate cancer are still largely missing from the picture when it comes to realisation. The vision has only moved to the clinical space in a small portion of medicine's disease panorama, and even a small portion of the cancer panorama.

Another related reason why the "PDTO for PO" vision does not seem hyped is that it almost never refers to preventive medicine. Predicting the future and preventing disease comes with more and other challenges than predicting what is needed in the present for a person with established disease (Green and Vogt 2016).

While the plausibility of today's promises is a theoretical question, the actual future of organoids is an empirical question to be documented.

4.4.2   Utility in Context

We have assessed the clinical utility of PDTOs as tools by considering not only scientific and theoretical, but technical, practical feasibility, and social limitations. This topic includes effectiveness, efficacy, cost-effectiveness, and safety.

We have also pointed at several hurdles regarding the practical feasibility of the vision. These relate to success rates in establishing organoids, generating enough organoids per patient, their speed of growth and timeliness of the testing, pain, clinical hazards, funding and gaining the trust of clinicians. Most importantly, we have pointed out a critical problem on technical variability between the models resulting from lack of standardisation and fundamental biological variability from self-organisation.

In terms of evidence, we have seen how the evidence on clinical utility of these organoids is still quite meagre. The largest, and most rigorous interventional study so far (Ooft et al. 2021), has a negative or inconclusive result and illustrates both practical issues with the strategy and the fact that making something work in socially embedded clinical practice is something very different from testing if the models are representative and predictive.

Importantly, only the patient-specific approach to delivery of PM through organoids has published evidence so far, probably because the biobanks for the stratified approach are too immature. We conclude that the utility of these models faces concrete challenges.

4.4.3 Problems of Knowing and Documenting that It Works

The clinical utility of PDOs for cancer treatment is still poorly documented, but there is a consensus in the field that observational studies have documented their clinical validity. Although we do not deny the ability of these models to be predictive, we have also challenged this assertion due to high risk of publication bias, and—again—limited research outside gastrointestinal cancer.

However, we can see more evidence coming in the "translational pipeline", including several randomised Phase II or Phase III trials. These will shed light on the relevance of the patient-specific approach in the coming years, but the stratified approach is largely missing in this pipeline. It will take years before robust clinical evidence for relevance is in place for some diseases, and many more until other diseases have been studied and the stratified approach has been properly tested.

Standardisation on various levels, in particular the process of developing organoids, is crucial for the reliability of results, being able to compare between clinics and labs, for the generalisability of findings (do they apply in other places than the particular clinic?) and for the trust of research subjects who expect that findings will be useful for other patients. At the same time, we have noted how standardisation comes into conflict with the very goal of personalised medicine.

Meanwhile, the nature of personalised medicine and the small groups of patients in each disease category, often does not permit larger studies. In this situation, single-case design studies—and their aggregates—may be one way forward. We have seen how a significant proportion of the clinically relevant studies are SCDs. Notably, none of these are rigorous n-of-1 trials but unblinded and non-randomised single patient open trials. Sometimes these are more rigorous, but often they take the form of anecdotes or simple case histories. This illustrates that n-of-1 trials may often be impractical or unethical in the organoid field, e.g. regarding cancer. However, it also illustrates that work needs to be done in developing and standardising single-case design methods in the field. This might fruitfully be done in collaboration with other such efforts (see e.g., Nikles et al. 2021). Finally, however, n-of-1 situations offer an evidence conundrum that will also affect the use of PDTOs. In many instances it will be hard to do costly and very rigorous n-of-1 trials, and one will have to handle uncertainty that cannot be quantified and therefore brings the clinician into the realm of the Art of medicine with qualitative, expert-based judgement.

4.5 Limitations of Our Method

We cannot—and have had little ambition of—predicting the future in this book. This is a limitation of all vision assessments, but it is also one of the strengths of its rationale: We aim to critique some of the building blocks of the argument for the vision in present, because we find it important to scrutinise the basis for future priorities and actions.

Our vision assessment has focused on scientific and theoretical limitations, and not wider considerations such as "why do we even envision and yearn for personalisation and a cure for cancer at all?" Compared to the way vision assessment has previously been framed, we have a quite narrow medical and philosophy of science-inspired gaze. We see this as a limitation, but also a methodological development of vision assessment.

Our material should not be seen as complete. There may be publications we have overlooked in our search for "review material," as well as published and registered research. There are certainly many key and interesting people we could have interviewed. In doing an amended HTA that looks beyond certain clinical studies, it is not easy to set the boundaries for or define the material one wants to look for.

It should be noted that the assessment conducted here is based on a study conducted at a given point in time, a snapshot of an emerging technology. The evidence base and realisation of the vision is evolving. However, the epistemological issues and major trends that we delineate is likely to develop slowly, and so many of our conclusions will remain valid for years. Some of them are even fundamental and will always be limitations to the field.

Finally, there is always a risk in vision assessments of degrading visions too much. Visions—or even hype—is not necessarily a bad thing. Visions are a critically important part of human creative endeavours, which the fields of organoids and their related technologies certainly is. At the same time, scientific visions for the future of medicine do not live in a scientific vacuum, but can also influence political prioritisations regarding funding, regulatory treatment and safety guidelines, as well as patients' expectations regarding biomedical technologies. For these reasons, we find it important to unpack the challenges that must be overcome to realise the vision. This book takes a step in this direction.

References

Acklin-Wehnert S et al (2023) Feasibility of establishing and drug screening patient-derived rectal organoid models from pretreatment rectal cancer biopsies. J Clin Oncol 41(4). https://doi.org/10.1200/JCO.2023.41.4_suppl.176

Baker L et al (2016) modelling pancreatic cancer with organoids. Trends Cancer 2(4):176–190. https://doi.org/10.1016/j.trecan.2016.03.004

Beckmann JS, Lew D (2016) Reconciling evidence-based medicine and precision medicine in the era of big data: challenges and opportunities. Genome Med 8(1):134. https://doi.org/10.1186/s13073-016-0388-7

Bengtsson A et al (2021) Organoid technology for personalized pancreatic cancer therapy. Cell Oncol 44:251–260. https://doi.org/10.1007/s13402-021-00585-1

Boers S et al (2016) Organoid biobanking: identifying the ethics. EMBO Rep 17:7

Boniolo G (2017) Molecular medicine: The clinical method enters the lab. What tumor heterogeneity and primary tumor culture teach us. In Philosophy of molecular medicine. Foundational issues in research and practice, eds. Marco Nathan, and Giovanni Boniolo, 23–42. New York and London: Routledge.

Bose S, Clevers H, Shen X (2021) Promises and challenges of organoid-guided precision medicine. Medicine 2(9):1011–1026. https://doi.org/10.1016/j.medj.2021.08.005

Bose S et al (2022) A path to translation: how 3D patient tumor avatars enable next generation precision oncology. Cancer Cell 40. https://doi.org/10.1016/j.ccell.2022.09.017

Brazovskaja A et al (2019) High-throughput single-cell transcriptomics on organoids. Curr Opin Biotechnol 55:167–171

Caiazza C, Parisi S, Caiazzo M (2021) Liver organoids: updates on disease modelling and biomedical applications. Biology 10:835. https://doi.org/10.3390/biology10090835

Chen H et al (2022) Urological cancer organoids, patients' avatars for precision medicine: past, present and future. Cell Biosci 12:132. https://doi.org/10.1186/s13578-022-00866-8

Chumduri C & Turco M (2021) Organoids of the female reproductive tract. Journal of Molecular Medicine 99:531–553 https://doi.org/10.1007/s00109-020-02028-0

Clara-Trujillo S et al (2020) In vitro modelling of non-solid tumors: how far can tissue engineering go? Int J Mol Sci 21:5747. https://doi.org/10.3390/ijms21165747

Clark J et al (2022) Novel ex vivo models of epithelial ovarian cancer: the future of biomarker and therapeutic research. Front Oncol 12:837233. https://doi.org/10.3389/fonc.2022.837233

de Vries YA et al (2018) The cumulative effect of reporting and citation biases on the apparent efficacy of treatments: the case of depression. Psychol Med 1–3. https://doi.org/10.1017/S0033291718001873

de Witte CJ, Valle-Inclan JE, Hami N, Lõhmussaar K, Kopper O, Vreuls CPH, Jonges GN, van Diest P, Nguyen L, Clevers H, Kloosterman WP, Stelloo E (2020) Patient-derived ovarian cancer organoids mimic clinical response and exhibit heterogeneous inter-and intrapatient drug responses. Cell Rep 31(11).

Djulbegovic et al., 2011Djulbegovic B, Hozo I, Greenland S (2011) Uncertainty in clinical medicine. In: Gifford F (ed) Philosophy of medicine, vol 16. Amsterdam, Elsevier

Engel GL (1977) The need for a new medical model: a challenge for biomedicine. Science 196(4286):129–136

Foo MA et al (2022) Clinical translation of patient-derived tumour organoids- bottlenecks and strategies. Biomark Res 10:10. https://doi.org/10.1186/s40364-022-00356-6

Francies H, Garnett M (2015) What role could organoids play in the personalization of cancer treatment? Pharmacogenomics 16(14):1523–1526

Green S, Vogt H (2016) Personalizing medicine: disease prevention in silico and in socio. Humana. Mente 30:105–145

Green S, Dam MS, Svendsen MN (2022) Patient-derived organoids in precision oncology – towards a science of and for the individual? In: Beneduce C, Bertolaso M (eds) Personalized medicine in the making. Series: human perspectives in health sciences and technology, vol 3. Springer, Cham, pp 125–146

Gkretsi V, Stylianopoulos T (2018) Cell adhesion and matrix stiffness: coordinating cancer cell invasion and metastasis. Front Oncol 8:145

Guillen K et al (2022) A human breast cancer-derived xenograft and organoid platform for drug discovery and precision oncology. Nature Cancer 3(2):232–250 https://doi.org/10.1038/s43018-022-00337-6

Günther C et al (2022) Organoids in gastrointestinal diseases: from experimental models to clinical translation. Gut 71(9):1892–1908. https://doi.org/10.1136/gutjnl-2021-326560

Hofer M, Lutolf M (2021) Engineering organoids. Nat Rev Mater 6:402–420. https://doi.org/10.1038/s41578-021-00279-y

Hoeyer K (2019) Data as promise: reconfiguring Danish public health through personalized medicine. Soc Stud Sci 49(4):531–555

Ishihara S, Haga H (2022) Matrix stiffness contributes to cancer progression by regulating transcription factors. Cancers 14(4):1049

Kane PB, Bittlinger M, Kimmelman J (2021) Individualized therapy trials: navigating patient care, research goals and ethics. Nat Med 27(10):1679–1686. https://doi.org/10.1038/s41591-021-01519-y

References

Kiwaki T, Kataoka H (2022) Patient-derived organoids of colorectal cancer: a useful tool for personalized medicine. J Pers Med 12:695 https://doi.org/10.3390/jpm12050695

LeSavage B et al (2022) Next-generation cancer organoids. Nat Mater 21:143–159. https://doi.org/10.1038/s41563-021-01057-5

Letai A, Bhola P, Welm A (2022) Functional precision oncology: testing tumors with drugs to identify vulnerabilities and novel combinations. Cancer Cell 40. https://doi.org/10.1016/j.ccell.2021.12.004

Li L, Selaru F (2022) Patient-derived functional organoids as a personalized approach for drug screening against hepatobiliary cancers. Adv Cancer Res 156(11). https://doi.org/10.1016/bs.acr.2022.01.011

Lin Y et al (2022) Progress and perspective of organoid technology in cancer-related translational medicine. Biomed Pharmacother 149:112869. https://doi.org/10.1016/j.biopha.2022.112869

Lo YH et al (2020) Applications of organoids for cancer biology and precision medicine. Nat Cancer 1:761–773. https://doi.org/10.1038/s43018-020-0102-y

Loong H et al (2020) Patient-derived tumor organoid predicts drugs response in glioblastoma A step forward in personalized cancer therapy. J Clin Neurosci 78:400–402. https://doi.org/10.1016/j.jocn.2020.04.107

Luo L et al (2022) Application progress of organoids in colorectal cancer. Front Cell Dev Biol 10:815067. https://doi.org/10.3389/fcell.2022.815067

Ma YS et al (2021) The power and the promise of organoid models for cancer precision medicine with next-generation functional diagnostics and pharmaceutical exploitation. Transl Oncol 14:101126

Mackenzie N et al (2022) Modelling the tumor immune microenvironment for precision immunotherapy. Clin Transl Immunol 11:e1400. https://doi.org/10.1002/cti2.1400

McEwen BS, Getz L (2012) Lifetime experiences, the brain and personalized medicine: an integrative perspective. Metabolism. https://doi.org/10.1016/j.metabol.2012.08.020

Meier MA et al (2022) Patient-derived tumor organoids for personalized medicine in a patient with rare hepatocellular carcinoma with neuroendocrine differentiation: a case report. Commun Med 2:80. https://doi.org/10.1038/s43856-022-00150-3

Moreno A, Ruiz-Mirazo K, Barandiaran X (2011) The impact of the paradigm of complexity on the foundational frameworks of biology and cognitive science. In: Hooker C (ed) Philosophy of complex systems. Elsevier, Amsterdam, pp 313–333

Mo S, Tang P, Luo W, Zhang L, Li Y, Hu X, Ma X, Chen Y, Bao Y, He X, Fu G, Hua G (2022) Patient-derived organoids from colorectal cancer with paired liver metastasis reveal tumor heterogeneity and predict response to chemotherapy. Adv Sci 9(31):2204097

Narashiman V et al (2020) Medium-throughput drug screening of patient-derived organoids from colorectal peritoneal metastases to direct personalized therapy. Clin Cancer Res 26(14):3662–3670

Nikles J et al (2021) Establishment of an international collaborative network for N-of-1 trials and single-case designs. Contemp Clin Trials Commun 23:100826. https://doi.org/10.1016/j.conctc.2021.100826

Ooft S et al (2021) Prospective experimental treatment of colorectal cancer patients based on organoid drug responses. ESMO Open 6(3):100103. https://doi.org/10.1016/j.esmoop.2021.100103

Pamarthy S, Sabaawy H (2021) Patient derived organoids in prostate cancer: improving therapeutic efficacy in precision medicine. Mol Cancer 20:125. https://doi.org/10.1186/s12943-021-01426-3

Pang MJ et al (2021) Gastric organoids: progress and remaining challenges. Cell Mol Gastroenterol Hepatol https://doi.org/10.1016/j.jcmgh.2021.09.005

Podaza E et al (2022) Next generation patient derived tumor organoids. Transl Res. https://doi.org/10.1016/j.trsl.2022.08.003

Poletti M et al (2020) Organoid-based models to study the role of host-microbiota interactions. J Crohn's Colitis 1–14

Qu J et al (2021) Tumor organoids: synergistic applications, current challenges, and future prospects in cancer therapy? Cancer Commun 41(12):1331–1353. https://doi.org/10.1002/cac2.12224

Reed M et al (2021) A functional precision medicine pipeline combines comparative transcriptomics and tumor organoid modelling to identify bespoke treatment strategies. Cells 10(12):3400. https://doi.org/10.3390/cells10123400

Rossi R et al (2022) Lung cancer organoids: the rough path to personalized medicine. Cancers 14:3703. https://doi.org/10.3390/cancers14153703

Schmäche T, Fohgrub J, Klimova A, Laaber K, Drukewitz S, Merboth F, Hennig A, Seidlitz T, Herbst F, Baenke F, Ada AM, Stange DE (2024) Stratifying esophago-gastric cancer treatment using a patient-derived organoid-based threshold. Mol Cancer 23(1):10

Schork N (2015) Personalized medicine: time for one-person trials. Nature 520:609–611. https://doi.org/10.1038/520609a

Schutgens F, Clevers H (2020) Human organoids: tools for understanding biology and treating diseases. Annu Rev Pathol 15:211–234. https://doi.org/10.1146/annurev-pathmechdis012419-032611

Shariati L et al (2021) Organoid technology: current standing and future perspectives. Stem Cells 1–25. https://doi.org/10.1002/stem.3379

Shiihara M, Furukawa T (2022) Application of patient-derived cancer organoids to personalized medicine. J Pers Med 12:789. https://doi.org/10.3390/jpm12050789

Sun CP et al (2022) Organoid Models for Precision Cancer Immunotherapy. Front. Immunol. 13:770465. https://doi.org/10.3389/fimmu.2022.770465

Sun Y, Ding Q (2017) Genome engineering of stem cell organoids for disease modelling. Protein Cell 8(5):315–327. https://doi.org/10.1007/s13238-016-0368-0

Tonelli MR, Shirts BH (2017) Knowledge for precision medicine: mechanistic reasoning and methodological pluralism. JAMA 318(17):1649–1650

Van den Berg A et al (2019) Personalised organs-on-chips: functional testing for precision medicine. Lab Chip 19(2):198–205. https://doi.org/10.1039/c8lc00827b

Verduin M et al (2021) Patient-derived cancer organoids as predictors of treatment response. Front Oncol 11:641980. https://doi.org/10.3389/fonc.2021.641980

Veninga V, Voest E (2021) Tumor organoids: opportunities and challenges to guide precision medicine. Cancer Cell 39. https://doi.org/10.1016/j.ccell.2021.07.020

Vogt H, Hofmann B (2022) How precision medicine changes medical epistemology: a formative case from Norway. J Eval Clin Pract 28(6):1205–1212

Vogt H (2025) Personalized medicine beyond stratification. In: Schramme T, Walker MJ (eds), Handbook of the philosophy of medicine, 2nd edn. Springer. In press

Walker M, Bourke J, Hutchison K (2019) Evidence for personalised medicine: mechanisms, correlation, and new kinds of black box. Theor Med Bioeth 40:103–121. https://doi.org/10.1007/s11017-019-09482-z

Wang J, Chen C et al (2022a) Patient-derived tumor organoids: new progress and opportunities to facilitate precision cancer immunotherapy. Front Oncol 12:872531. https://doi.org/10.3389/fonc.2022.872531

Wang Q, Guo F et al (2022b) Applications of human organoids in the personalized treatment for digestive diseases. Signal Transduct Target Ther 7:336. https://doi.org/10.1038/s41392-022-01194-6

Wang Y, Li Y et al (2022c) Advances of patient-derived organoids in personalized radiotherapy. Front Oncol 12:888416. https://doi.org/10.3389/fonc.2022.888416

Wensink G et al (2021) Patient-derived organoids as a predictive biomarker for treatment response in cancer patients. Precis Oncol 5:30. https://doi.org/10.1038/s41698-021-00168-1

Woodcock J, Marks P (2019) Drug regulation in the era of individualized therapies. N Engl J Med 381:1678–1680. https://doi.org/10.1056/NEJMe1911295

Wu HY et al (2022) Correlation between drug sensitivity profiles of circulating tumour cell-derived organoids and clinical treatment response. Eur J Cancer 166:208–218. https://doi.org/10.1016/j.ejca.2022.01.030

References

Xu H et al (2022) Tumor organoids: applications in cancer modelling and potentials in precision medicine. J Hematol Oncol 15:58 https://doi.org/10.1186/s13045-022-01278-4

Yan Y et al (2022) Patient-derived rectal cancer organoids—applications in basic and translational cancer research. Front Oncol 12:922430. https://doi.org/10.3389/fonc.2022.922430

Zhang M et al (2020) Generation of 3D human gastrointestinal organoids: principle and applications. Cell Regen 9:6. https://doi.org/10.1186/s13619-020-00040-w

Zhou Z et al (2021) Patient-derived organoids in precision medicine: drug screening, organoid-on-a-chip and living organoid biobank. Front Oncol 11:762184. https://doi.org/10.3389/fonc.2021.762184

Open Access This chapter is licensed under the terms of the Creative Commons Attribution 4.0 International License (http://creativecommons.org/licenses/by/4.0/), which permits use, sharing, adaptation, distribution and reproduction in any medium or format, as long as you give appropriate credit to the original author(s) and the source, provide a link to the Creative Commons license and indicate if changes were made.

The images or other third party material in this chapter are included in the chapter's Creative Commons license, unless indicated otherwise in a credit line to the material. If material is not included in the chapter's Creative Commons license and your intended use is not permitted by statutory regulation or exceeds the permitted use, you will need to obtain permission directly from the copyright holder.

Chapter 5
Vision Management: Towards Responsible Visioneering for Tumour Organoid Technology in Precision Oncology

5.1 Summary of Results

In Chapter 2, we described visions as prospects guiding scientific developments and their early application in the clinic. Visions and the way they are conveyed by researchers and technology developers motivate research efforts and justify the translational endeavour. This is particularly relevant for biomedical research. In that sense, visions impact many stakeholders: the general public and specifically patients who are or will be directly concerned as possible benefiters of the technology (or being harmed from it); clinicians in search for new treatments and options for their patients; policymakers who are looking for best strategies for orienting, regulating, and funding research; the industry and private investors eager to lead the development of potentially profitable emerging technologies and services.

Our assessment focused on the main, overarching promise about patient-derived tumour organoids (PDTOs) improving predictions of what the patient needs and providing significant clinical utility to individuals through personalisation of cancer treatment. The degree to which this promise seems credible or plausible based on our evaluation of its foundations, depends on the boldness in which the promise is made. There are promises of revolutions, but there are mostly more modest promises of incremental gains and specific applications and successes. We have noted several problems related to how representative and reliable the models are and can become, as well as uncertainties related to how they should be developed, their practical use and their documentation through scientific methods. However, it is important to note that models need not be perfect to be useful. Simple models with a specific use may still represent a gain in the clinic.

The degree of hype – or the distance between what is promised and how far the field has progressed—is also different in different areas of the field. The actual focus of existing clinical studies and the current "translational pipeline" of registered trials is on certain types of cancer (especially gastrointestinal), while little progress has been made with regard to other types. We have also pointed out that all the published and registered clinical research is on the patient-specific approach of delivering personalised medicine, and not the stratified approach, which is based on large "living biobanks" of organoids that have not yet become mature enough for this use.

Several hundreds of cancer patients have had clinical outcomes correlated to or predicted by organoid results when we collected our material and that number is rising. However, although we have uncovered an emerging body of interventional clinical research testing the actual use of the models to predict what treatments patients should have, this research cannot document clinical utility at this point. There is a larger body of clinically relevant observational research that merely documents that the organoids react in a similar way as patients (or their tumours) to the same treatments and that this can predict patient responses. Although this may look promising at this point, we have questioned the validity of this picture by pointing out a high risk of publication bias.

5.2 Does the Community Correctly Know the Evidence? Addressing Publication Bias

As has been shown in Chapter 4, the exercise in assessing how plausible the vision is, has turned out to be difficult because of the scarcity of evidence and the lack of common standards among publications. Publication bias is one of the main concerns that has emerged during our inquiry. Although there are reviews of the literature that make an effort to compile published evidence, it is difficult, even for agents in the field, to have a clear representation of the status of evidence.

Importantly, there is potentially a strong publication bias in the published literature. A first clue towards this conclusion is that it is difficult to systematically match completed clinical trials registered in international databases to resulting publications. Considering that not all clinically relevant experiments might be registered in trials databases, it seems that a lot of clinically relevant information does not make it through the public domain. The lack of publication of negative results might result in a bias in the data available—and this holds also for all results, positive or negative, not published because of small patients' groups. In the end, studies that show a positive correlation between PDOs and patients are likely to be overrepresented in the published data. In other words, that would mean that PDTOs are probably a less reliable prediction tool for the individual than all the evidence available suggests.

The likely distortion is problematic for assessing the actual evidence status, especially in the context of health technology assessment. A clear and explicit strategy should be developed to address publication bias in the field. Individual researchers and groups have a responsibility, along with institutions (research performing organisations, research funding organisations, publishers…) who have a duty to make it easy for researchers to publish all relevant results, so that personal incentives and collective interests align. While a general strategy to address publication bias through, e.g., best practice scientific data management (Wilkinson et al. 2016) is often claimed for, there are also domain-specific issues. Precision medicine favouring single-case design trials faces the challenge of assessing the evidence from a few patients or even an individual. In general, emerging technologies in need of favourable results to maintain funding and investment may be particularly prone to publication bias.

5.3 How Long Until Clinical Application? Striving for Standardisation

An important prerequisite for clinical application that emerged almost as an unconsciously shared motto across our interviews of different experts, is standardisation. Standardisation has two sides. It is first an important aspect of documentation and knowledge production. At the exploration stage, all technical options should be pursued. But if procedures are not commensurable from one laboratory to another, testing a hypothesis against reliable and reproducible data is impossible. Standardisation is also a prerequisite for clinical applications, if the technology is to be delivered cost-effectively and routinely in the clinic. At some point, these two requirements of standardisation merge. Before entering clinical practice, the product must be approved by regulatory authorities based on the documentation of its efficiency. As one of our interviewees said:

> I think we are in a proof-of-concept phase, where organoids are at a large extent basic research, where publications are coming out, we are exploring, we are finding correlation studies and show we could have been predictive, or we could be predictive but on anecdotal basis, small sample size, not representative of the field or the market or the incidents. And we are not very close to any prototyping phase. I think there are a lot of proof-of-concepts, but when I am saying prototype: it is a clinical prototype, it is standardized, it is something that you can go to the FDA with and say: 'we know exactly what is in this and we know with a 95% certainty with this error margin that we can get that result'. We are far away from that type of prototype. And only when we achieve that, we will be able to get clinical trials approved, for market approval, for precision medicine care. (senior researcher, industry)

The challenge of standardisation is part of a process that goes until industrial production and a shared methodology to deliver the benefits of the technology in the clinic. Standardisation, as a step in this general process of clinical translation, is not only a technical aspect of the delivery of the promises of biomedical research. There is also an ethical responsibility of researchers and research organisations to produce reproducible science, science that can contribute to the overall effort of research, especially when publicly funded and using patient material.

That does not mean that research should be limited in scope or channelled through a standardisation pipeline that would be imposed from above or from the mood of the momentum. Basic research must explore all possible technical options, even the less dominant ones, for we do not know what is going to emerge from this exploration process. However, when entering the clinical domain, and if a specific promise is made to patients and other stakeholders that clinical delivery is the horizon, then a serious effort needs to be made in the direction of shared products and methodologies and systematic measures of outcomes.

Finally, as our analysis of published evidence and registered trials suggested, research is conducted worldwide. This internationalisation of research, in the context of scarce evidence and few studies, makes the importance of standardisation and avoiding publication bias even more flagrant. When possible, it would be beneficial for all the research and health community to discuss shared standards and ethical issues at the global level.

5.4 Who Would Benefit? Addressing Economics and Distributive Justice Issues

The public is concerned that organoids as emerging technologies are not pursued at the cost of increasing healthcare inequalities (Ravn et al. 2023). This calls for the consideration of two aspects: the cost of the technology and the population of patients who are likely to benefit from it.

Will most concerned patients be able to benefit from the technology? When the technology turns to clinical practice, an obstacle might be its cost and the availability of the technology. Our interviewees shared some concerns about these two points. The cost of the technology might not make it available to everyone, especially in healthcare systems that rely mostly on private payment and insurance. On a similar note, some researchers-clinicians among our interviewees were concerned that they are now providing their patients with PDTOs tests relying on research money. Given that research funding is limited in time and scope, there might be no resources for offering the test to other patients in the absence of a clinical trial.

It seems that there are too many unknowns on the determination of the cost of an individual PDTO test. Now conducted in research laboratories, the current labour-intensive and exploratory methods will have to be replaced by procedures automated or routinely executable by laboratory technicians. These technological developments depend on the ability to scale down the price of the test in near future. Another practical issue for the access of the technology is the option of conducting the test onsite or sending the cells to a dedicated platform. In the context of PM, genetic tests have been standardised and commercialised through the circulation of samples and data, sometimes internationally. What would be needed to realise PDTOs testing beyond research settings? All the practical issues, related to rigorous evaluations of cost and cost-efficiency of specific tools, would have to be discussed at a later stage, once the technology closes into the clinic and regulatory approval. At the vision stage, it is difficult to provide more than general comments.

In this regard, it is interesting to note the strengths and also limits suggested by the antibiogram analogy (i.e., a PDTO test for cancer would be similar to an antibiotic resistance test for a given bacteria, see Chapter 4). The analogy points to a device that is not available yet. When claiming that, in the future, a PDTO test should be

5.4 Who Would Benefit? Addressing Economics and Distributive Justice Issues

as simple as an antibiogram, interviewees just set a horizon and highlight that we are not there yet. But more interestingly, the analogy echoes also the fact that not all bacterial infections got cured thanks to an antibiotic that has been chosen on the basis of an antibiogram. Most of the time, a probabilistic approach is enough to determine the right antibiotic, and only severe cases make it to the antibiogram testing. The same will be probably true of PDTOs for PO.

If, I, as a patient, hear the promise that clinicians are now able to "find the right drug for the right patient" thanks to the new PM techniques, I might expect my organoids to be grown and used to find the best therapy from the start. However, from the viewpoint of the healthcare system and the practitioner, this might still mean that the standard EBM treatment will be given priority, and only then, when no results are obtained, genetic screening, then functional screening (i.e., PDTOs) might be considered (see e.g., Green et al. 2022). In this process, only a small subset of cancer patients will be concerned by the tool. In terms of distributive justice: How many patients can the technology finally reach and how many benefits in terms of quality-adjusted life years (QALY) can we really expect to justify the investment it requires to reach the clinical practice?

There are probably some specific subpopulations who can expect a significant gain from PDTOs, such as patients engaged in Phase I trials.

> If you are going to try to figure out when patients are going to benefit, it is in the early phase development – the Phase I of a drug. You are going to preselect [patients] based on organoid responses, and there you have immediate benefits from patients that are done with their standard of care treatment, because they have a better patient selection for enrolling in Phase I studies, which is now being done unselected, which is ridiculous. (junior researcher, industry)

How significant the gain might be, this subpopulation still represents a very small portion of the people potentially concerned.

This leads us to the second question, which is the overall cost-effectiveness of the technology. The overall cost to society is not only the cost of the diagnostic test paid by the patient or the hospital, but more generally the amount of R&D investment put in the technology. If we look at the general cost of the biotechnology all along its lifecycle, there are several points that deserve attention in the perspective of building a responsible vision.

A strong argument in favour of PDTOs as a strategy to identify the right treatment is the potential improvement in the diagnostic and prognostic as a way to improve the cost-effectiveness of cancer treatment. Screening is about sparing fruitless efforts, that is, therapies not efficient for a given individual and that represent unnecessary suffering and costs.

> Maybe doing a diagnostic test for prediction based on organoid is much more expensive than sequencing DNA. Still, it is so much cheaper than giving an expensive therapy to a patient who will not benefit from that. I see more the healthcare system to explode due to the price of the therapies than the price of the diagnostic tests. (senior researcher, academic)

> Very expensive diagnostic organoid testing upfront saves us from doing a very expensive surgery which a lot of complications that in the end [patients] don't need, or having drugs that are extremely expensive over many years. (senior researcher, academic)

The population that may benefit from the technology might also evolve in the future. Even if we do not aim at producing a tumour antibiogram for each patient, precision medicine could benefit from the stratified and the biobank approach.

> To me, the concept of precision medicine should be that: over time, in this learning system – so it is not just a system where you treat somebody and then that's it, but you use that data and hopefully you will use the information for future treatment – within this learning system that over time you have more effective healthcare. It is not going to happen in two years, it is going to take a decade or more, and maybe 20 years is more realistic, I don't know. (senior researcher, academic)

The path to industrial development is paved with many unknowns as well. Profitability still relies on an uncertain economic model. PDTOs are just a test, and after a test comes the treatment: Is the test provider and the treatment provider different companies? Which market mechanisms ensure that all stakeholders can benefit from their investment and that patients can benefit from the innovation? Making this model cost-effective and available to all assumes that issues on ownership of the biobank, potential patent trolling, market mechanisms, are discussed at some point and resolved in the sense of the general interest.

5.5 How Should One Communicate the Vision? Dealing with Hype

It is generally deemed important that researchers should refrain from exaggerating the prospects of the technology, i.e., resorting to "hype" (Huch et al. 2017). The demand of sharing publicly a sober vision, with realistic and specific prospects, is based on the prerequisite of honesty (exaggeration might distort the true representation of the field), on the fact that exaggerating prospects can incline patients and other stakeholders to misplaced hope, and that disappointed hope or even illusion might in turn lead to a process of loss of trust in science and in the discourse of scientists.

As noted above, the answer to the question of how much the technology is hyped depends on the identification of a specific vision. It is important in this regard that all stakeholders delineate as clearly as possible the vision defended and communicated. There are different degrees of boldness in the vision, and there are also different specific fields of applications that deserve to be articulated to avoid misunderstandings. Obviously, relying on patient-derived organoids for individual treatment screening is not the same thing as developing organoids in a tissue engineering perspective for transplants. And yet, many published reviews and editorials make loosely the distinction between different applications. This blurs the vision and undermine the credibility of the promises in the long run. What is the potential spectrum of diseases to be covered by the technology: Are all cancer types a target on the same footage? There is a contrast between reviews that claim clinical prospects for organoids in many domains of cancer, and the reality of published evidence focusing on some types. This "extension challenge"—how far a methodology for which evidence is piling up regarding some types of cancer can be extended to other types—should be recognised as such in public statements.

Our interviewees, who are key actors of the field, took different perspectives when questioned about hype. Visioneering is part of scientific research and researchers cannot be blamed for sharing a vision or acting towards its realisation. But some recognise that organoid technology developers themselves might have been too optimistic when it comes to clinical applications.

Hype is not only exaggerating prospects towards the general public and patients. There is also a source of hype when fundamental assumptions are made implicit and when hidden difficulties are not reported explicitly and highlighted in scientific articles. For instance, the practical difficulties to grow organoids (success rate, time, labour required…) are hurdles towards an efficient process and the delivery of clinical applications. Not reporting the "learning curve" might incline the reader, even a researcher, scientist or clinician, to expect more practical outcomes of the technology than it ought. As one of our interviewees stated:

> If the casual scientific reader looks at some of these papers, they would say 'oh, you know, I can just tomorrow start growing organoids, let me follow these methods and I'll grow organoids.' It is much more challenging than that and I think somehow in the articles they rarely define the challenges (senior researcher, academic).

A first step towards a more responsible communication, even in the scientific community itself, is to be explicit about these limitations and hurdles, especially in terms of practical feasibility. Also, one should keep in mind that standards for clinical delivery are higher than those for proof-of-concept.

A role can be attributed to publishers and conditions of publications as well. In original studies, requiring a full reporting of methodology and data, including some clinical aspects that usually remain in shadow because of the lack of concrete endpoints, would contribute to highlight the difficulties and represent a first effort towards standardisation. Data availability might naturally counter hype, because some limits of the technology or its amenability will be made more evident. In discussion papers, it is important to make room for the challenges as well, including practical ones. Also, discussion papers could be framed to provide different perspectives on the prospects of a technology, including a critical stance.

The management of expectations is a collective endeavour. Our technological future is not only shaped by opportunities and limitations of the technology itself, but also in the path that we embrace when investing our money or our hopes in one direction over another. Although they have already been tested, patient derived organoids for precision medicine are still more of a prospect than a reality. With our exercise in assessing the prospects and current challenges of this emerging technology, we have tried to contribute to defining the right level of expectations for all stakeholders, from policymakers to patients.

References

Green S, Dam MS, Svendsen MN (2022) Patient-derived organoids in precision oncology – towards a science of and for the individual? In: Beneduce C, Bertolaso M (eds) Personalized medicine in the making. Series: Human Perspectives in Health Sciences and Technology, vol 3. Cham, Springer, pp 125–46

Huch M et al (2017) The hope and the hype of organoid research. Development 144(6):938–941

Ravn T, Sørensen MP, Capulli E, Kavouras P, Pegoraro R, Picozzi M, Saugstrup LI, Spyrakou E, Stavridi V (2023) Public perceptions and expectations: disentangling the hope and hype of organoid research. Stem Cell Reports 18(4):841–852

Wilkinson MD, Dumontier M et al (2016) The FAIR guiding principles for scientific data management and stewardship. Sci Data. 3:160018. https://doi.org/10.1038/sdata.2016.18

Open Access This chapter is licensed under the terms of the Creative Commons Attribution 4.0 International License (http://creativecommons.org/licenses/by/4.0/), which permits use, sharing, adaptation, distribution and reproduction in any medium or format, as long as you give appropriate credit to the original author(s) and the source, provide a link to the Creative Commons license and indicate if changes were made.

The images or other third party material in this chapter are included in the chapter's Creative Commons license, unless indicated otherwise in a credit line to the material. If material is not included in the chapter's Creative Commons license and your intended use is not permitted by statutory regulation or exceeds the permitted use, you will need to obtain permission directly from the copyright holder.

Chapter 6
Summary and Concluding Remarks

Our aim in this book has been twofold: First, we have aimed to develop a method for the assessment of something that is at once as powerful and as intangible as a vision, which has not yet been realised in practice. Going beyond the framework of a traditional health technology assessment (HTA), we have further developed and expanded the methodological framework of a vision assessment for this purpose. Second, we have applied this methodology to the specific vision of using organoids as a tool to realise precision medicine, as patient-specific models to guide treatment selection in the realm of cancer medicine.

This book was motivated by the observation that visions for new medical technologies can have profound, unintended, and unforeseen consequences on different levels. Through political and economic support, visions can drive scientific developments and public expectations of medicine. Thus, in addition to an evaluation of existing evidence of benefits and risks (the tasks of a traditional HTA), we propose the framework of a vision assessment examining also the epistemological assumptions, prerequisites, and conditions for the vision of interest. Building on previous scholarship elaborating on the traditional HTA framework, our proposed vision assessment builds on four types of written material: scientific review papers and other publications from the field outlining the vision, interviews from informants at the cutting edge of clinical translation (epistemic hotspots), clinically relevant published evidence, and clinically relevant registered trials. This allows us to identify the main expectations in relation not only to existing evidence, but also to reflect on the extent to which the evidence that is "in the pipeline" can be expected to address remaining challenges in the field.

We have applied this methodological framework in examining the implications of the vision of using organoids as patient-specific models in precision medicine. We have focused on the field closest to clinical implementation, namely patient-derived tumour organoids (PDTOs) in cancer medicine. This envisioned application raises intriguing philosophical questions about what constitutes evidence in the era of precision medicine, aiming to address existing shortcomings of EBM and forecast what works in the individual case. We have critically analysed the epistemic assumptions underlying the main expectations: that PDTOs more accurately capture the complexity and variation of a patient's specific tumour, that implementation of the models is feasible in the clinic, and that their clinical utility can be assessed through acceptable methods and evidence standards.

We found that PDTO presents a promising avenue for "functional precision medicine", where organoids "stand in" for the patient as a form of "avatar" or biological replica to yield make reliable, yet molecularly agnostic, predictions. But although these models are made from the cancer patient's own tumour cells, their trustworthiness for clinical decision making cannot just be taken for granted, as they are still relatively simple models. Published observational trials have been conducted to document the predictive potential of PDTOs, but evidence documenting the clinical utility through interventional trials is still lacking. One reason is due to the early stage of the emerging technology, where RTCs are being planned but have not yet been completed. But other important reasons can be unpacked through an examination of the challenges in early interventionist trials and revealed through interviews with people at the forefront of this field.

Among the identified challenges are epistemic uncertainties stemming from mismatches between the complexity of the tumour microenvironment versus the in vitro culture, lacking standardisation of protocols and procedures, as well as practical challenges relating to the time-frame for growing organoids for real-time testing for clinical purposes. Interviewed informants also raised concerns about publication bias, where negative results from early trials are not reported. Although preclinical and clinical models need not be perfect to be useful, the existing challenges call for moderation of expectations to organoids as available tools for patient-specific testing. Moreover, we have pointed to unsolved issues of affordability and distributive justice as organoids, and PM in general, may be accessible only to resourceful patients. Still, organoid research presents a promising avenue for biomedical development, as there is an increasing and pressing challenge to identify new ways of predicting what works in an individual patient.

Our vision assessment not only point to strategies of vision management for stakeholders but also towards new research avenues for philosophers of science. As highlighted throughout the book, organoid research provides fascinating insights to current redefinitions of evidence in medicine with the increasing emphasis on personalised treatment opportunities. While policy reports often give the impression that PM will provide the right treatment for the right patient—and at the right time—this broader vision is in practice challenging to achieve. Organoid research is both a response to challenges in PM and an illustration of the reconfiguration of evidence. On one hand, organoids assays may hold a solution to the limitations of attempts to predict treatment response from genomic analysis alone. On the other hand, phenotypic screening using organoids also raises questions about whether the complexity of cancer can be sufficiently recapitulated in vitro. Giving persistent uncertainties about this, new questions are raised about what constitutes evidence when models

6 Summary and Concluding Remarks 107

become patient-specific and sample sizes converge towards one. These will be important questions to explore in future philosophical studies, and we hope that this book can inspire readers embarking on this journey.

Open Access This chapter is licensed under the terms of the Creative Commons Attribution 4.0 International License (http://creativecommons.org/licenses/by/4.0/), which permits use, sharing, adaptation, distribution and reproduction in any medium or format, as long as you give appropriate credit to the original author(s) and the source, provide a link to the Creative Commons license and indicate if changes were made.

The images or other third party material in this chapter are included in the chapter's Creative Commons license, unless indicated otherwise in a credit line to the material. If material is not included in the chapter's Creative Commons license and your intended use is not permitted by statutory regulation or exceeds the permitted use, you will need to obtain permission directly from the copyright holder.

Appendices

Appendix A: Published clinically relevant evidence

*By "number of organoid/patient correlations established" we mean the following: the number of patients included in the study is relevant for assessing the practical feasibility of PDOs, but is not an absolute indicator of the power of the study, which is first limited by the organoid growth success rate. Indeed, of 2 studies including the same number of patients, one may have much more organoids established than the other (of note, the number of patients included is not always communicated). Then, some organoid data might not be readable, or researchers might not be able to compare organoid data with patient's data (e.g., because patients dropped out). That's why we deem that the most relevant indicator of the quantity of evidence provided by a study is the total number of patients whose organoids have been successfully established and for which organoid data has been parallelled with clinical data.

Study: publication year and title	Types of disease/ organoid	Study design	Number of organoid/ patient correlations established*
Hill et al. (2018). Prediction of DNA Repair Inhibitor Response in Short Term Patient-Derived Ovarian Cancer Organoids	High grade serous ovarian cancers (HGSC)	Observational	22
Puca et al. (2018). Patient derived organoids to model rare prostate cancer phenotypes	Neuroendocrine prostate cancer	Observational	4

(continued)

(continued)

Study: publication year and title	Types of disease/ organoid	Study design	Number of organoid/ patient correlations established*
Sachs et al. (2018). A Living Biobank of Breast Cancer Organoids Captures Disease Heterogeneity	Breast, primary, metastatic, or recurrent tumour sites	Observational	12
Tiriac et al. (2018). Organoid profiling identifies common responders to chemotherapy in pancreatic cancer	Pancreatic, from primary tumours (hT) and metastases (hM)	Observational, retrospective study	8
Vlachogiannis et al. (2018). Patient-derived organoids model treatment response of metastatic gastrointestinal cancers	Metastatic gastrointestinal (colorectal + gastroesophagal)	Observational, prospective	8
Yan et al. (2018). A Comprehensive Human Gastric Cancer Organoid Biobank Captures Tumour Subtype Heterogeneity and Enables Therapeutic Screening	Gastric cancoocer (primary)	Observational	3
Beltran et al. (2019). A Phase II trial of the aurora kinase A inhibitor alisertib for patients with castration resistant and neuroendocrine prostate cancer	Prostate cancer, neuroendocrine	Observational	2
Driehuis et al. (2019). Pancreatic cancer organoids recapitulate disease and allow personalised drug screening	Pancreatic	Observational	4

(continued)

Appendices

(continued)

Study: publication year and title	Types of disease/ organoid	Study design	Number of organoid/ patient correlations established*
Driehuis et al. (2019). Oral Mucosal Organoids as a Potential Platform for Personalised Cancer Therapy	HNSCC (head and neck)	Observational	7
Ganesh et al. (2019). A rectal cancer organoid platform to study individual responses to chemoradiation	Rectal, primary, metastatic, or recurrent disease	Observational	26
Ooft et al. (2019). Patient-derived organoids can predict response to chemotherapy in metastatic colorectal cancer patients	Metastatic colorectal cancer	Prospective, observational multicenter clinical study	29
Pasch et al. (2019). Patient-Derived Cancer Organoid Cultures to Predict Sensitivity to Chemotherapy and Radiation	Metastatic colorectal	Observational, prospective	1
Phan et al. (2019). A simple high-throughput approach identifies actionable drug sensitivities in patient-derived tumour organoids	Ovarian and high-grade serous tumours	Observational	2
Steele et al. (2019). An Organoid-Based Preclinical Model of Human Gastric Cancer	Gastric	Observational	2
Votanopoulos et al. (2019). Model of Patient-Specific Immune-Enhanced Organoids for Immunotherapy Screening- Feasibility Study	Melanoma stage III-IV	Observational	7

(continued)

(continued)

Study: publication year and title	Types of disease/ organoid	Study design	Number of organoid/ patient correlations established*
Arena et al. (2020). A Subset of Colorectal Cancers with Cross-Sensitivity to Olaparib and Oxaliplatin	Metastatic colorectal cancer	Observational	3
Derouet et al. (2020). Towards personalised induction therapy for esophageal adenocarcinoma-organoids derived from endoscopic biopsy recapitulate the pre-treatment tumour	Esophageal adenocarcinoma	Observational	5
Frappart et al. (2020). Pancreatic cancer-derived organoids—a disease modelling tool to predict drug response	Pancreatic, metastatic	Observational	1
Jacob et al. (2020). A Patient-Derived Glioblastoma Organoid Model and Biobank Recapitulates Inter- and Intra-tumoral Heterogeneity	Glioblastoma	Observational	7
Janakiraman et al. (2020). Modelling rectal cancer to advance neoadjuvant precision therapy	Rectal, stage II/III	Observational	5
Jiang et al. (2020). An Automated Organoid Platform with Inter-organoid Homogeneity and Inter-patient Heterogeneity	Gastrointestinal, rectal, and liver	Observational	3

(continued)

Appendices

(continued)

Study: publication year and title	Types of disease/ organoid	Study design	Number of organoid/ patient correlations established*
Li et al. (2020). Rapid screening for individualized chemotherapy optimization of colorectal cancer	CRC (potentially operable)	Observational, "double-blind co-clinical cohort study"	17
Loong et al. (2020). Patient-derived tumour organoid predicts drugs response in glioblastoma. A step forward in personalised cancer therapy	Glioblastoma	Interventional	1
Narasimhan et al. (2020). Medium-throughput Drug Screening of Patient-derived Organoids from Colorectal Peritoneal Metastases to Direct Personalised Therapy	Colorectal cancer (CRC) patients with peritoneal metastases	Interventional element in prospective, observational study	19
Sharick et al. (2020). Metabolic Heterogeneity in Patient Tumour-Derived Organoids by Primary Site and Drug Treatment	Pancreatic cancer	Observational	7
Xu et al. (2020). Patient-derived organoids in cellulosic sponge model chemotherapy response of metastatic colorectal cancer	Metastatic colorectal cancer	Observational	12
Yao et al. (2020). Patient-Derived Organoids Predict Chemoradiation Responses of Locally Advanced Rectal Cancer	Rectal, locally advanced rectal cancer (LARC)	Prospective, observational (within Phase III trial)	80

(continued)

(continued)

Study: publication year and title	Types of disease/ organoid	Study design	Number of organoid/ patient correlations established*
Beutel et al. (2021). A Prospective Feasibility Trial to Challenge Patient–Derived Pancreatic Cancer Organoids in Predicting Treatment Response	Pancreatic	Observational, prospective	16
Bi et al. (2021). Successful Patient-Derived Organoid Culture of Gynaecologic Cancers for Disease Modelling and Drug Sensitivity Testing	Endometrial carcinoma, HER2 positive	Observational	1
Cho et al. (2021). Patient-derived organoids as a preclinical platform for precision medicine in colorectal cancer	Colorectal, mostly stage III or IV CRC	Observational	40
De Witte et al. (2021). Patient-Derived Ovarian Cancer Organoids Mimic Clinical Response and Exhibit Heterogeneous Inter- and Intrapatient Drug Responses	Ovarian	Observational	5
Forsythe et al. (2021). Organoid Platform in Preclinical Investigation of Personalised Immunotherapy Efficacy in Appendiceal Cancer-Feasibility Study	Appendiceal cancer	Observational	2

(continued)

(continued)

Study: publication year and title	Types of disease/ organoid	Study design	Number of organoid/ patient correlations established*
Guillen et al. (2021). A breast cancer patient-derived xenograft and organoid platform for drug discovery and precision oncology	Breast cancer with early metastatic recurrence	Observation and intervention	1
Karakasheva et al. (2021). Patient-derived organoids as a platform for modelling a patient's response to chemoradiotherapy in oesophageal cancer	Oesophageal cancer	Prospective, observational	4
Kim et al. (2021). Modelling Clinical Responses to Targeted Therapies by Patient-Derived Organoids of Advanced Lung Adenocarcinoma	Lung cancer (adenocarcinoma)	Observational	9
Ooft et al. (2021). Prospective experimental treatment of colorectal cancer patients based on organoid drug responses	Metastatic colorectal cancer (CRC)	Prospective interventional trial, single center	19
Pan et al. (2021). Breast cancer organoids from malignant pleural effusion-derived tumour cells as an individualized medicine platform	Breast cancer	Observational	1

(continued)

(continued)

Study: publication year and title	Types of disease/organoid	Study design	Number of organoid/patient correlations established*
Park et al. (2021). A Patient-Derived Organoid-Based Radiosensitivity Model for the Prediction of Radiation Responses in Patients with Rectal Cancer	Locally advanced rectal cancer	Prospective observational	19
Reed et al. (2021). A Functional Precision Medicine Pipeline Combines Comparative Transcriptomics and Tumour Organoid Modelling to Identify Bespoke Treatment Strategies	Glioblastoma, rare disease	Interventional/observation	1
Wang et al. (2021). Accuracy of Using a Patient-Derived Tumour Organoid Culture Model to Predict the Response to Chemotherapy Regimens In Stage IV Colorectal Cancer	Stage IV Colorectal Cancer, metastatis sampled in several	Prospective observational, blinded	45
Wang et al. (2021). Conversion Therapy of Intrahepatic Cholangiocarcinoma Is Associated with Improved Prognosis and Verified by a Case of Patient-Derived Organoid	Intrahepatic cholangiocarcinoma	Observational	1
Cao et al. (2022). Patient-Derived Organoid Facilitating Personalised Medicine in Gastrointestinal Stromal Tumour With Liver Metastasis—A Case Report	Gastrointestinal stromal tumours (GIST)	Observational	1

(continued)

(continued)

Study: publication year and title	Types of disease/ organoid	Study design	Number of organoid/ patient correlations established*
Chen et al. (2022). Patient-derived tumour organoids as a platform of precision treatment for malignant brain tumours	Malignant brain tumours	3 interventional case studies (as part of larger study)	3
Meier et al. (2022). Patient-derived tumour organoids for personalised medicine in a patient with rare hepatocellular carcinoma with neuroendocrine differentiation case report	Rare hepatocellular carcinoma	Interventional, case report	1
Seppälä et al. (2022). Precision Medicine in Pancreatic Cancer-Patient-Derived Organoid Pharmacotyping Is a Predictive Biomarker of Clinical Treatment Response	Pancreatic Cancer	Prospective study leveraging data from a randomised controlled clinical	12
Wu et al. (2022). Correlation between drug sensitivity profiles of circulating tumour cell-derived organoids and clinical treatment response	Pancreatic ductal adenocarcinoma (PDAC), circulating tumour cells (CTCs)	Observational, prospective cohort (one case report seems interventional)	31
Yao et al. (2022). Application of tumoroids derived from advanced colorectal cancer patients to predict individual response to chemotherapy	Advanced colorectal	Prospective, observational	34

Appendix B: Overview of registered interventional trials

Trial number	Disease	Trial type (Phase)	Patients (N)	Trial description	Year/status	Country
NCT03544047	Breast cancer	Unspecified (exploratory)	50	A single-arm exploratory clinical study to evaluate the consistency and accuracy of PDO models of breast cancer to predict the clinical efficacy of the drug, as well as the possibility of guiding the neoadjuvant chemotherapy	January 2019- July 2020 (status unknown)	China
NCT05136014	Lung cancer	Unspecified (experimental for non-responders)	200	This trial is registered as an observational study to evaluate the effects of tyrosine kinase inhibitors via PDO models, but organoid-based results will inform clinical decision making for non-responders	November 2021 (status unknown)	France
NCT03500068	Pancreatic cancer	Unspecified	30	Clinical trial (phase not specified) involving the use of PDOs to assess efficacy of first-line treatment as well as guide second-line treatment options for metastatic pancreatic cancer patient	September 2017- September 2022 (still recruiting)	The Netherlands
NCT04279509	Refractory solid tumours	Phase I	35	Phase I trial using PDOs to conduct a 10-drug panel screening to inform treatment decisions for patients with refractory solid tumours (HNSCC, Ovarian cancer, Colorectal cancer)	May 2019 – May 2022 (status unknown)	Singapore
NCT04727632	Breast cancer	Phase I (early stage)	6	Early Phase I trial, a companion to the IRB # 131,027 FORESEE trial comparing imaging to drug profiling results from PDOs	March 2021 - April 2025	USA
NCT05177432	Breast cancer	Phase I	26	Phase I, non-randomised, study including patient-specific testing of 10–12 anti-cancer drugs on PDOs	December 2021- December 2025	Singapore
NCT05432518	Brain cancer	Phase I (early stage)	10	Early Phase 1 trial, single-arm open-label prospective study with PDOs used to decide between 5 treatments for patients with recurrent glioblastoma	September 2022 -December 2027	Canada

(continued)

(continued)

Trial number	Disease	Trial type (Phase)	Patients (N)	Trial description	Year/status	Country
NCT05532397	Brain cancer	Phase I	10	Phase I, non-randomised, study including PDO-screening for patients with astrocytic glioma who have exhausted other treatment options	November 2022 -December 2025	Singapore
EU CTR -2014–003,811-13	Colorectal cancer and lung cancer	Phase I	61	Phase I, non-randomised, study including including PDO-screening for patients with only one or no standard treatment options	April 2016 – April 2019 (completed)	The Netherlands
NCT04450706	Breast cancer	Phase I	15	Phase I, non-randomised, study including PDO-screening for patients who have exhausted other treatment options	August 2023 -August 2025	USA
NCT05381038	Breast and gastric cancer	Phase I Phase II	10	A combined Phase I and Phase II trial, where participants identified to potentially benefit from PDO-selected treatment (Phase I) will transition to a dose modulation phase (Phase II) to evaluate the most efficient drug dosis	June 2022 – April 2027	Singapore
NCT05078866	Lynch Syndrome and related diseases (colorectal cancer)	Phase I Phase II	45	A combined Phase I and II trial where patient-specific colorectal adenoma organoids are used to evaluate the safety and effect of the Nous-209 vaccine against colorectal cancer in Lynch syndrome patients	September 2022 - July 2025	Collaboration between 3 cancer centers in the USA and a university in Puerto Rico
NCT05024734	Bladder cancer	Phase II (non-randomised)	33	Phase II, non-randomised ang single-group assignment, involving treatment selection based on PDO-testing of four drugs. The drug with the highest antitumour effect on PDOs will be chosen for patients	November 2022 – November 2024	Switzerland

(continued)

Appendices

(continued)

Trial number	Disease	Trial type (Phase)	Patients (N)	Trial description	Year/status	Country
NCT05464082	Breast cancer	Phase II (non-randomised)	80	Phase II, non-randomised, study using both PDX and PDO models to prospectively evaluate the correlation between engraftment success with recurrence. Upon recurrence, the patient-derived models will inform treatment selection	December 2021- September 2027 (suspended, accrual pending completion of amendment)	USA
NCT05267912	Advanced colorectal cancer (CRC) and advanced solid cancers	Phase I (pilot-study) Phase II (randomised)	1919	ORGANOTREAT TRIAL: A combined study involving a pilot-study restricted to advanced CRC (Phase I) and two Phase II studies in advanced solid cancers. A tumour board will make treatment recommendations based on a PDO-based chemogram report. One Phase II study is single arm (2A), the second is randomised to compare the efficacy of chemogram-driven treatment versus standard of care. A cross-over will allow patients in the control arm to benefit from chemogram-based treatments. [14]	January 2019- January 2026	France
NCT05669586	Lung cancer	Phase II (randomised)	50	Phase II study (randomised) enrolling lung cancer patients who are resistant to multi-line standard therapies. Patients are randomised into groups where choice of antitumour therapy is guided by i) physicians or ii) PDO-based drug sensitivity test (intervention)	March 2023 – May 2026	China

(continued)

(continued)

Trial number	Disease	Trial type (Phase)	Patients (N)	Trial description	Year/status	Country
EU CTR 2020–003,395-41	Colorectal cancer	Phase II (single arm)	45	Phase II, open-label single arm, to investigate the efficacy of pharmacogenomic profiling and patient-specific drug sensitivity testing on metastatic colorectal cancer biopsies. Response rates (complete and partial) will be assessed for the total population and each study drug cohort	September 2020 (ongoing)	Norway
NCT05378048	Abdominal cancer	Phase II (randomised)	140	Phase II proof-of-concept randomised controlled trial in patients with inoperable or metastatic abdominal tumours to use PDOs to guide choice of chemotherapy. The study involves comparison of PDO drug response to patients' clinical response	July 2022 – July 2025	China
NCT05429684	Breast cancer	Phase III (non-randomsed)	120	Phase III trial with parallel assignment, non-randomised. Patients and paired PDO models are divided into six groups according to genomic signatures. Each patient group will receive the estimated best targeted treatment scheme, while the matched PDO model will accept a variety of potential effective schemes intervention. The future treatment plan of patients will be adjusted based on the tumour inhibition rate of PDO models	January 2021 - February 2024	China

(continued)

Appendices

(continued)

Trial number	Disease	Trial type (Phase)	Patients (N)	Trial description	Year/status	Country
NCT05351398	Gastric cancer	Phase III (randomised)	54	The phase is not specified in trial description, but the design study design is an RCT with two groups of patients with stage III gastric cancer divided into: i) An intervention group where treatment selection for patients needing neoadjuvant chemotherapy is guided by PDO-based drug sensitivity assay and ii) a traditional group receiving standard treatment allocation	April 2022-December 2023	China
NCT04842006	Colorectal cancer	Phase III (randomised)	93	Phase III trial (RCT) aiming to reduce overtreatment of patients that most likely will not benefit from additional treatments. Patients with high-risk features will be randomised to i) a treatment strategy with early systemic control by chemotherapy followed by circulating tumour DNA (ctDNA) and organoid-guided adjuvant therapy, or ii) to conventional treatment strategy	December 2021-December 2031	Finland
NCT04931394	Pancreatic cancer	Phase III (randomised)	200	Phase III trial (RCT) to explore whether chemotherapy regimens guided by organoid drug sensitivity testing can improve the outcomes of advanced pancreatic cancer. The study will evaluate establishment rate of organoids from biopsy tissue and concordance between drug sensitivity test results and patients' treatment response	June 2021-May 2025	China

Index

A
Access to healthcare, 102
Amenability, 17, 44, 69, 103
Antibiogram, 40, 42, 85, 100–102
Anticipation, 4, 12
Assumptions, 1, 7, 8, 14, 15, 17, 18, 25, 29, 30, 36, 38, 58, 59, 71, 85, 86, 88
Avatar, 39–42, 47, 54, 55, 57, 58, 65–67, 73, 84, 85, 106
Avatar models, 40

C
Communication, 13–17, 102–103
Credibility, 15, 88, 102
Culture media, 55, 63

D
Design principles, 38, 54, 58, 63, 88

E
Emerging technologies, 1, 2, 7, 8, 13, 14, 16–18, 23, 91, 97, 99, 100, 103, 106
Epistemological hotspot, 19, 21, 26, 30, 32, 67, 73–75, 80–82
Epistemological uncertainty, 4, 5, 8, 15, 16, 24, 44, 53, 84
Evidence-based medicine (EBM), 5, 29, 45, 67, 84, 86, 87
Evidence production, 45, 47, 69

F
Feasibility, 16, 24
Foresight, 12

Functional precision medicine, 33, 38, 39, 106, 116

H
Health technology assessment (HTA), 5, 7, 11–14, 16, 24, 70, 91, 98, 105
Hype, 7, 12, 13, 17, 53, 88, 91, 97, 102, 103

I
Intratumor heterogeneity, 66

L
Living biobank, 30, 41, 43, 79, 98, 110

M
Management of expectations, 103
Microenvironment, 37, 56–58, 60, 83, 106
Model uncertainty, 1, 16, 53, 88
Molecular agnosticism, 38, 80

N
N=1, 6, 20, 39, 73, 86, 87

O
Organoid ethics, 4
Organ-on-a-chip (OoC), 3, 19, 20, 23, 58, 61, 88

P
Patient-derived organoid (PDO), 3, 34, 36, 37, 39, 42, 43, 45, 65, 82, 83, 85,

102, 110, 111, 113, 115, 116, 119–123
Patient-derived tumour organoids (PDTOs), 3, 4, 16, 29, 30, 32, 34, 70, 72, 80, 84, 85, 87, 89, 97, 100, 105, 106, 111, 113, 117
Patient-specific approach, 39, 40, 46, 62, 68, 70, 74, 75, 79, 89, 90, 98
Personalised medicine (PM), 2, 4, 20, 29, 34, 37, 39, 82, 83, 85, 90, 98, 117
Practical feasibility, 1, 16, 18, 38, 43, 62, 65, 73, 83, 89, 103, 109
Precision medicine (PM), 2, 5–7, 20, 29, 32, 34, 35, 57, 62, 80, 83, 87, 99, 102, 103, 105, 106, 114, 117
Precision oncology, 3–5, 8, 29, 30, 32, 115
Predictive models, 3, 36, 86
Predictive models, 3, 36, 86
Publication bias, 1, 74, 79, 86, 90, 98–100, 106

R
Representation, representativeness, 16, 23, 35, 36, 38, 42, 47, 53–56, 59–61, 85, 86, 88, 98, 102

S
Self-organization, 38, 58, 63, 64, 88, 89
Sociology of expectations, 12, 13
Standardization, 55, 62, 64, 69, 74, 79–83, 86, 89, 90, 99, 100, 103, 106
Stem cells, 2–4, 31, 32, 35, 36, 38, 55, 56, 58, 64
Stratified approach, 39, 41, 43, 45, 62, 68–70, 79, 89, 90, 98

T
Technology assessment (TA), 7, 11, 13, 14, 18
Translational pipeline, 16, 24, 26, 75, 78–81, 90, 97

V
Vision analysis, 14, 15, 20, 21, 25, 29, 30, 53, 59, 62, 68, 70, 72, 84, 87, 88
Vision assessment, 1, 8, 13–15, 17–19, 21, 24, 88, 90, 91, 105, 106
Visioneering, 14, 103
Vision evaluation, 15, 16, 20, 21, 25, 26, 30, 40, 42, 47, 61, 75, 87
Vision management, 15, 16, 18, 106

The manufacturer's authorised representative in the EU is Springer Nature Customer Service Centre GmbH, Europaplatz 3, 69115 Heidelberg, Germany. If you have any concerns regarding our products, please contact ProductSafety@springernature.com

Printed and bound by CPI Group (UK) Ltd, Croydon, CR0 4YY
26/03/2026
02078992-0004